'Timed to influence the 2021 UN climate cl
Mackintosh presents a compelling case for ui
the challenge of our era. Dispensing acute a
highly readable manifesto for engagement s.
globalization 2.0. This book is an indispensable guide for business, educators,
journalists, government officials and every citizen. It is a major achievement made
possible by Mackintosh's decades-long global experience in the European Union,
with Mitsubishi, and as executive director of the Group of Thirty and as president
of the National Association of Business Economists.'

Dr. Carl Lankowski, *retired director of European Area Studies,*
US Department of State

'Dr. Mackintosh has written a long overdue book. Climate Crises Economics
covers this vital subject fully in a candid and unusually complete manner. His book
can be read by the expert or educated layman. One of his skills is weaving together
a number of recommended solutions rather than one draconian one. An important
read for those who care, a must-read for those who want to put a solution in place.'

Michael O. Clark, *Senior Advisor at 1919 Investment Counsel, LLC, USA*

'Few areas of study combine the efforts of science, economics and public policy as
does climate change analysis. Stuart Mackintosh, with his background in economics
and public policy, is excellently positioned to combine the threads of these three
efforts into a manuscript that describes an approach to achieve the goals of public
and private agents in the field of climate change. I endorse this book as a must-read
for those who wish to actively pursue the goals of addressing climate change in a
systematic way.'

John Silvia, *CEO and Founder of Dynamic Economic Strategy, USA*

CLIMATE CRISIS ECONOMICS

Climate Crisis Economics draws on economics, political economy, scientific literature, and data to gauge the extent to which our various communities – political, economic, business – are making the essential leap to a new narrative and policy approach that will accelerate us towards the necessary transition to a decarbonized economy and sustainable future.

The book draws out policies and practices with both national and local examples, which will demonstrate various complementary approaches that are empowering states and people as they seek to pursue the carbon neutral goal. The author delineates a climate crisis economics approach that is fit for purpose and which can help achieve necessary climate change goals in the decades ahead. Ensuring economic and ecological sustainability is neither easy nor cost-free; there is no single solution to the climate crisis. All aspects of our economies, policies, business, and personal practices must come into alignment in order to succeed. Frustratingly, we know what is needed and we have many of the technologies and systems to make the leap to a carbon neutral economy, yet we still fail to act with alacrity. Leaders, communities, and businesses must shift their narratives in how they talk about and think about the climate crisis. In doing so, in making the narrative leap to a new understanding about what is possible and necessary, we can stop endangering our common future and single, fragile, global habitat, and instead set the stage for Green Globalization 2.0 and a new, sustainable industrial revolution.

Climate Crisis Economics will appeal to academics, students, investors, and professionals from varying disciplines including politics, international political economy, and international economics. Written in an accessible voice, it draws on work in fields outside of and in addition to politics and economics to make a case for climate crisis economics as an approach to addressing the climate change challenge ahead.

Stuart P. M. Mackintosh is Executive Director of the Group of Thirty (www.g30.org), an influential international economic and financial think tank comprised of the most senior figures in central banking, finance, and academia. In 2016 he was elected by his peers as President of the National Association of Business Economics, the leading US organization of professional economists. He is also a Visiting Fellow at Newcastle University, UK. He has a deep and broad international professional network across the US and globally and he is the author of *The Redesign of the Global Financial Architecture* (2016; 2nd ed. 2021). www.stuartmackintoshauthor.com

CLIMATE CRISIS ECONOMICS

Stuart P. M. Mackintosh

LONDON AND NEW YORK

First published 2022
by Routledge
2 Park Square, Milton Park, Abingdon, Oxon OX14 4RN

and by Routledge
605 Third Avenue, New York, NY 10158

Routledge is an imprint of the Taylor & Francis Group, an informa business

Cover image and design by Geoffrey Laurence. See http://geoffreylaurence.com/

British Library Cataloguing-in-Publication Data
A catalogue record for this book is available from the British Library

Library of Congress Cataloging-in-Publication Data
Names: Mackintosh, Stuart P. M., author.
Title: Climate crisis economics/Stuart P.M. Mackintosh.
Description: Milton Park, Abingdon, Oxon; New York, NY: Routledge, 2022. |
Includes bibliographical references and index.
Identifiers: LCCN 2021013974 (print) | LCCN 2021013975 (ebook) |
ISBN 9780367478704 (hardback) | ISBN 9780367478698 (paperback) |
ISBN 9781003037088 (ebook)
Subjects: LCSH: Climatic changes–Economic aspects. |
Climatic changes–Political aspects.
Classification: LCC QC903 .M32 2022 (print) |
LCC QC903 (ebook) | DDC 363.738/74561–dc23
LC record available at https://lccn.loc.gov/2021013974
LC ebook record available at https://lccn.loc.gov/2021013975

ISBN: 978-0-367-47870-4 (hbk)
ISBN: 978-0-367-47869-8 (pbk)
ISBN: 978-1-003-03708-8 (ebk)

DOI: 10.4324/9781003037088

Typeset in Bembo
by Newgen Publishing UK

For Una, John, and Bernard

CONTENTS

ILLUSTRATIONS

BOXES

ACKNOWLEDGEMENTS

I would like to thank all those who have supported me during my research for this book. My particular thanks to Don Swenholt, for keeping me on track and on time. Thanks also to Michael Clark, Carl Lankowski, John Silvia, and Brian Sturgess for their reaction to the proposal at the outset. Thanks also is due to Geoffrey Laurence, who conceived of, sketched, and designed the cover of the book.

I must also thank my wife, Jean, for allowing me the space to complete this endeavour and for her tolerance of far too many discussions of the subject, the process, and the difficulties it presented.

Writing a book is a solitary exercise, but it also impinges on much else around it, and I am appreciative of my friends and colleagues who have supported the process.

I would also thank my good friend and Diane Stamm for her careful eye as to style and presentation. I really valued her eye, and her willingness to say when more work was needed. Thanks also to my researcher Aubrey Byrum for her hard work.

Finally, I must thank Emily Ross and Hannah Rich, both superb editors, who shepherded the book from its raw form to completion.

Stuart P. M. Mackintosh
Washington DC, February 2021

ABBREVIATIONS

AI	artificial intelligence
BES	biodiversity and ecosystem services
CCC	Committee on Climate Change
CCS	carbon capture and storage
CEO	chief executive officer
CO_2	carbon dioxide
COP	Conference of the Parties
COP26	26th UN Climate Change Conference of the Parties
CORSIA	Carbon Offsetting and Reduction Scheme for International Aviation
Covid-19	Coronavirus disease 2019
DICE	Dynamic Integrated Climate-Economy model
ECB	European Central Bank
EIB	European Investment Bank
EPA	US Environmental Protection Agency
EPI	Economic Policy Institute
ESG	environmental, social, and governance
ETF	exchange traded fund
ETS	Emissions Trading System
EU	European Union
EV	electric vehicle
EV2G	EV-to-grid technology
GDP	gross domestic product
GHG	greenhouse gas
GND	Green New Deal
$GtCO_2e$	gigatons of CO_2 equivalent
GW	gigawatts

G20	Argentina, Australia, Brazil, Canada, China, France, Germany, India, Indonesia, Italy, Japan, Republic of Korea, Mexico, Russia, Saudi Arabia, South Africa, Turkey, UK, US, and EU
HCC	Haute Conseil pour le Climat
HES	*homo economicus sympatico*
IAMs	Integrated Assessment Models
IASB	International Accounting Standards Board
IEA	International Energy Agency
IFC	International Finance Corporation
IMO	International Maritime Organization
IPCC	Intergovernmental Panel on Climate Change
MDBs	multilateral development banks
MTCO$_2$	metric tons of CO_2
MVP	minimum viable product
NCB	National Carbon Bank
NGFS	Network for Greening the Financial System
NOAA	National Oceanic and Atmospheric Administration
OECD	Organisation for Economic Co-operation and Development
PCF	permafrost climate feedback
PE	private equity
PETM	Palaeocene–Eocene thermal maximum
PV	photovoltaic
R&D	research and development
RGGI	Regional Greenhouse Gas Initiative
SARS	severe acute respiratory syndrome
SDG	sustainable development goal
SO$_2$	sulphur dioxide
TCFD	Taskforce for Climate-Related Financial Disclosure
UN	United Nations
WCI	Western Climate Initiative
WCO	World Carbon Organization
WEF	World Economic Forum
WTO	World Trade Organization
XR	Extinction Rebellion
YTD	year-to-date

PROLOGUE

An imagined vision of the near future without action on climate change

The day is Saturday, January 1, 2050. Global temperatures have been increasing steadily and are now 2 degrees Celsius above pre-industrial levels and appear to be headed relentlessly upwards. On this New Year's Day, the CO_2 in the atmosphere is 505 parts per million, a level not seen in more than 4 million years. It is a grim climate way-post. Scientists have concluded that tipping points of no return are a climate and global certainty. In 2050, it is no longer a matter of whether a tipping point will be reached but which one is next, and how soon.

Catastrophe looms everywhere.

Summer Arctic sea ice disappeared in 2043. In response, the remaining oil majors made a final gamble. In a last-gasp dash for carbon fuels, they rushed into Arctic exploration, adding to the negative greenhouse gas (GHG) emissions dynamics and further undermining climate goals. The same damaging fossil fuel lobbies that had worked successfully in the three prior decades to undermine efforts to price and tax carbon are still at work. Their win is the planet's loss.

In 2050, almost all alpine glaciers are gone, save those in the Himalayas. Melting glaciers adversely impact water supplies for billions in India, China, and elsewhere, as the great rivers of the world slow and shrink.

Sea levels are rising at a rate of nearly 2 metres per century, double what had been expected, with scientists concerned that an event comparable to the Palaeocene–Eocene thermal maximum (PETM), or Younger Dryas climate shift events, could occur.

Severe flooding and the frequency and severity or hurricanes and typhoons have increased. Major cities around the globe – from Washington, DC, New Orleans, Manhattan, and Miami in the US to Ho Chi Minh City in Vietnam, Bangkok in Thailand, London in England, and Amsterdam in the Netherlands – are hit hard. The great cities and their peoples bear the rising cost of flooding, erosion, property damage, diminished health of their residents, and their economies amounting to

DOI: 10.4324/9781003037088-1

trillions of US dollars. Driven by climate change, these ocean and weather extremes call into question the viability of coastal cities. A growing movement in the US calls for the Capitol to be moved to Philadelphia.

Increasing levels of ice cap melt are reported in Greenland. The US Defense Department ran scenarios of a partial collapse of the ice cap and in response moved their physical assets away from the coasts.

The West Antarctic Ice Sheet is calving vast icebergs and its instability is increasing. Scientists warn that parts of the East Antarctic Ice Sheet are also showing signs of instability.

Each summer, large parts of Siberia are ablaze. The boreal forests suffer, and permafrost is melting. Summer temperatures commonly peak at over 38 degrees Celsius (100.4 degrees Fahrenheit) inside the Arctic Circle. Cargo ships report huge plumes of bubbling water in the Arctic, as vast stores of frozen methane hydrate melt and add to the climate feedback loops already playing out.

The death of the great Amazon rainforest is an increasing worry in 2050. Researchers have concluded that rising temperatures, shifting precipitation, and illegal deforestation compromise the capacity of the rainforest to act as a carbon sink.

The Anthropocene extinction event has been unfolding largely without action from states preoccupied with short-termism and nationalist backlash against collective, coordinated climate change solutions. Fully one-quarter of the world's species are on the brink of extinction or are already extinct, among them polar bears, the Great whales, the mountain gorillas of Rwanda, bluefin tuna, scores of shark species, numerous birds, and thousands of insect species. Australia's unique flora and fauna are among the hardest-hit ecosystems, with many marsupial species disappearing. An immense, ongoing tragedy of environmental and species destruction is playing out, unchecked. Some regions of the globe have tipped over into new climate change states, a new equilibrium.

Australia suffers huge annual wildfires, scorching millions of acres, while its interior bakes at temperatures as high as 50 degrees Celsius (122 degrees Fahrenheit), hot enough to kill a person caught outside during the day. Farming has been devastated and is economically unviable. Much of the country's interior is parched to desertification. Swiss Re's 2020 warning of ecosystem collapse is coming true.

By 2050, California has been suffering major wildfires year after year after year. Despite massive expenditure and a clear commitment to net zero, the state has been unable to arrest temperature trends driven by others. The wildfire season now extends through most of the year. Water shortages are mounting. The impact on air quality has been significant. This has filtered through to liveability, with more and more people moving away to locations as distant as Denver, Colorado, in search of cleaner air and relief from the smoke.

Governments and their leaders bear the bulk of the blame for these failures. Historians point to the 2021 COP26, and the years following the disappointment in Glasgow, as the pivotal period of political weakness and betrayal.

Timid leaders, afraid of angry voters still reeling from the devastation of the Covid-19 pandemic, were unwilling to make the necessary planetary and climate change narrative leaps. Leaders at COP26 failed to seize the opportunity to reset expectations across markets and economies on the climate change glidepath to net zero. In hindsight, COP26 was viewed by observers as the last real opportunity to secure a sustainable climate change narrative, a net-zero transition that could be achieved without significant economic disruption and societal unrest. Failure in Scotland contributed to paralysis at subsequent COP gatherings. And with each passing year, the glidepath to net zero steepened. Short-term economic costs of the transition rose, making the needed consensus and actions more difficult.

Although leaders at COP26 agreed carbon pricing was essential to the net-zero transition, the actual price levels, and a requirement for the prices to progressively rise towards a common, agreed goal, was not part of the final declaration. This politically courageous but necessary economic step was not taken at subsequent COP meetings, either. As a result, too many sectors and firms avoided paying the true planetary costs, and GHG emissions continued to rise.

Carbon taxes and cap-and-trade schemes did become more common in the years immediately after COP26, but they lacked an agreed level of pricing or yearly increases. Consequently, the overall levels of pricing of carbon remained too low to significantly affect incentives and market conduct in the way needed to achieve net-zero goals in the most polluting markets.

Some regions and countries, notably those in the European Union (EU), designed technocratic and legal enforcement of net-zero plans that worked, and bent their national GHG emission curves, almost reaching national net-zero goals by 2050. Yet, absent similar schemes in every country, Europe's impact on GHG emissions was not enough to affect the overall rise in emissions by free-rider states, unregulated markets, and polluting firms.

For the brief period of 2021–2024, the Biden administration attempted a leap to a new climate change narrative and policy stance through, among other things, the US Green New Deal. This was welcomed globally. However, domestic US political constraints meant pricing, regulatory shifts, and the Green New Deal were stymied. Democrat losses in the 2022 midterm congressional elections further slowed US policy action. A victory by the Republican Party, under a new Trumpian leader in 2024, reversed this modest progress on climate change policy, and America turned its back on climate change mitigation, the transition, and the planet. Once again, nativist nationalist politics precluded collective action and actively undermined common climate goals. Geopolitical tensions rose.

Tensions between the two remaining superpowers, the US and China, mounted in the 2020s. The US was backsliding and again becoming isolationist and denialist on climate change. China sped forward in its climate change transition, aiming to seize a dominant position in green economy technologies. As the US declined, China grew faster and greener, as did Europe. This divergence – denialism versus green dynamism – caused a split, in the mid-2020s, pitting the US against a new alliance of China, the EU, and a few other key states on climate change policy and response. Although the new so-called 'red-green-blue' China–EU axis did go on to make large strides in their net-zero goals, it proved impossible to secure global emissions goals because of the US reversal.

Leaders missed the opportunity to change the carbon narrative. Crucially, there was a lack of coordinated international enforcement of commitments. The United Nations (UN) did try to oversee (or at least report) on carbon pricing. However, as with so many prior attempts, the organization was unable to enforce commitments and convert them into implementation. The UN showed that, outside geopolitical diplomacy, it could not operate as an enforcement body because member nations made no commitment to enforcement.

As a result, in the 2020s and beyond, markets were not incentivized effectively or overseen strictly on GHG emissions. Under-supervised and under-regulated, markets and firms did not fully internalize the cost of carbon. Many firms adhered to the letter of the law (such as it was) but not to the spirit of national regulations, too often resulting in little or no GHG reductions. Markets continued to operate without assurances that firms that made measurable transitions to net-zero processes and strategies would be rewarded and those that did not would be penalized. The stock of GHG mounted and the GHG flows continued. Here again, some states, primarily in Europe, took more robust regulatory action, and their firms made the leap to the climate change net-zero narrative. But without broader international application and compliance, the effect was not enough to shift global market sentiments and actions.

Left to their own devices, new markets for offsets, carbon derivatives, and securities were created, but these were gamed by both sellers and buyers. The sellers sold worthless securities that made no difference to GHG emissions, and the buyers purchased them to comply with requirements, knowing (or certainly suspecting) their inherent worthlessness. Money was made, wasted, diverted, and the planet warmed.

Some markets, industries, and firms did respond in 2021 and in the decades after Glasgow to demands that they shift their strategies and business narratives. Numerous leading firms and certain sectors made the transition into solar, wind, batteries, and electric vehicles and transport. These businesses did deliver for the planet and their consumers. Investors demanded change and they got it when some governments set the guardrails, expectations, and regulations, and the firms responded. Industrial sectors and firms, principally located in Europe, China, and other green economies, increased market share and outperformed the polluters. However, these successes were not matched in other economic sectors, and the net-zero transition faltered.

By 2050, too many actors and firms had failed to make a rapid, complete transition. Most notably, industrial manufacturing and construction lagged. Here, the systems and products (such as cement and metals manufacturing) failed to shift to net-zero circular economy processes, and the needed global retrofitting and rebuilding remain incomplete.

It is not only the industrial and construction sectors that are responsible for net-zero failures. Consumers and farmers are also to blame. Too many consumers continue to eat too much meat, predominantly beef. Agricultural practices have not changed in major markets, especially in China and the US. As a result, ruminant GHG emissions have not only not declined but have risen, as too few people have changed their diets and humanity eats their way towards the next climate tipping point.

Failure in Glasgow and since has also contributed to political and economic crises in weak states. In the almost 30 years since COP26, some poorer countries have been pushed to failure by climate change shocks and societal breakdown. From Central America to the Middle East, to Africa, and to parts of Asia, countries have been weakened by corruption, and by demographic and economic pressures, all exacerbated by ecosystem crises. These states have been unable to provide for their populations, and their people have done what they must: They have left in the tens and hundreds of millions. The climate change refugee migration crisis precipitated in the 2040s was orders of magnitude greater than that seen in the US and in Europe in the 2020s.

In the end, a failure by advanced economy leaders to commit to modest, manageable annual investments of funds and a 2 percent (of GDP) transfer of funds to support the green transition to net zero has hobbled countries and has contributed to geopolitical crises and instability. This foolhardy parsimony meant countries could not make the net-zero transformation. Instead, communities buckle under the weight of climate disasters and civil and societal unrest. The continuing migration crises fuel nationalism and an anti-outsider politics in advanced economy elections and politics, further undermining any possibility for climate change progress.

Looking back, the cost of global climate change mitigation is widely recognized as having been entirely manageable – a bargain – economically. But poor economic advice, poor model design, a lack of leadership, woeful policy implementation, selfishness, and short-sighted decisions in too many key states made the necessary leap and policy changes difficult for some and politically impossible for too many.

Greta Thunberg, who as a teenager had ignited a global movement with her visionary, courageous call to her elders to pay attention to climate change and to do something about it, 'right here, right now', was by 2050 the leader of climate activists globally. But she reacted to the continuing global intransigence to adequately address climate issues by finally resigning from public life, convinced she could not shift the narrative.

Is this imagined 2050 possible? Unfortunately, yes. We may fail at COP26 in Glasgow. We may fail to act collectively as we must to arrest climate change and assure planetary survival. But this dystopian planetary disaster is not yet inevitable. Though shrinking, a time frame still exists to act.

This book argues that if we learn lessons from the current Covid-19 and prior crises, and learn lessons about our models, pricing, incentives, institutions, markets, and policies, we can still shift the climate change narrative – i.e. our collective story and understanding about climate change – and accept our responsibility for a workable solution and take appropriate action. We can set regulatory and market expectations, harness markets, and secure an equitable, sustainable, resilient transformation to a net-zero economy before we trigger tipping points to a hothouse world of no return.

To do this, we need to make war on carbon. Achieving decarbonization of our economies requires changes across all our economies and societies. But leaders, businesses, and communities can make the net-zero transition. Some countries, sectors, leaders, firms, and communities are already doing so. If we make the leap, plan the way ahead, and lay out, monitor, and enforce policy implementation, the glidepath is still manageable.

Climate Crisis Economics argues the transition is still achievable and identifies the necessary models, policy, and practice to secure the transition.

1
CRISES AS CRUCIBLES FOR CHANGE

Right here. Right now. This is where we draw the line.

Thunberg, 2019

Net zero is not a slogan. It is a scientific imperative.

Carney, 2020

Politics must not be subject to the economy, nor should the economy be subject to the dictates of an efficiency-driven paradigm of technocracy. Today, in view of the common good, there is urgent need for politics and economics to enter a frank dialogue in the service of life, especially human life.

Francis, 2015: 189

The climate change crisis is the existential challenge of our time. The future of the planet and the survival of all living things depends on our collective effort to arrest global temperature increases, for that will determine the sustainability of all life. The relentless increase in atmospheric carbon dioxide (CO_2) is indisputable, disturbing, and increasingly damaging. Humanity's short-sightedness and selfishness, and our inability to raise our gaze to the climate risk horizon, to recognize the common good and the fragility of the planet, our only home, may yet doom us. Things cannot continue as they are. As Greta Thunberg said in her impassioned speech at the 2019 UN Climate Summit in New York, the time to act is 'right here, right now' (Thunberg, 2019).

CO_2 levels in the atmosphere stood at 414 parts per million in January 2021 (Figure 1.1), the highest level ever recorded in human history. Global temperatures reflected this dangerous fact; 2020 was the hottest year on record (*Scientific American*, 2020). The stifling heat of 2020 occurred without a major El Niño event such as the one that boosted global temperatures to a new high four years ago. Therefore, 2020

DOI: 10.4324/9781003037088-2

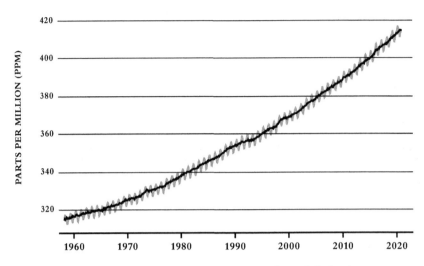

FIGURE 1.1 Relentless human-driven increase in atmospheric CO_2 levels

Source: Scripps Institution of Oceanography, NOAA Global Monitoring Laboratory, January 2021.

sent an ominous signal about continuing the long-term warming trend driven by human activities that emit GHGs.

Severe weather events linked to climate change underscored the effects of rising temperatures on the planet. Siberia, normally a vast frozen landscape, in 2020 had the hottest year on record, with some locations reaching over 37 degrees Celsius, another world record. This was a temperature that, absent human CO_2 emissions, would be expected to occur once every 80,000 years. Record and above-average temperatures were reported around the world. Fires ravaged California, scorched Australia, and blackened the Amazon jungle. Temperatures in Iraq hit an all-time record: You could cook a steak rare standing in the sunshine or overcook a good piece of salmon. Floods, hurricanes, and typhoons slammed coastlines across the globe. Over a third of Bangladesh was flooded by a typhoon. The US reported the largest number ever of named storms in one season – 30 – causing the National Oceanic and Atmospheric Administration (NOAA) to run out of alphabetical letters to name them and requiring Greek letters for the last hurricanes of the season. Before this jump, the norm would have been 12 named storms in a season (CBS, 2020).

Such repeated, relentless record-breaking numbers portend a terrible future climate for us all if we fail to act to address and pursue net-zero carbon emissions, which is a scientific imperative (Carney, 2020). Humanity's impact on the planet's temperature, the 'hockey stick' jump visible in long-run climate data, from ice cores, is firmly established and well known (Figure 1.2) (Littlemore, 2009). Our responsibility for the Anthropocene epoch[1] and the extinction event that we are collectively causing is clear and damning.

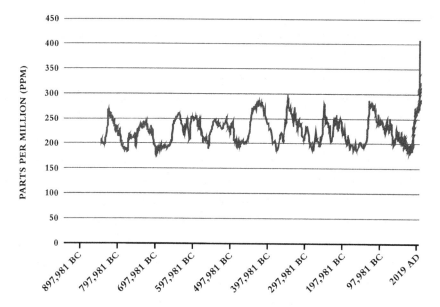

FIGURE 1.2 Atmospheric CO_2 levels at highest level in 800,000 years

Source: US National Oceanic and Atmospheric Administration (IS NOAA), 2018; G30, 2020.

Today, a series of looming, horrific, nonlinear cascading climate change tail risks – the loss of the Arctic sea ice, the melting of the Alpine glaciers, the collapse of the Greenland ice sheet, the dieback of the Amazon and boreal forests, the collapse of the East and West Antarctic ice sheets, the melting of Siberia's permafrost – should haunt our nightmares. Yet we draw these catastrophes ever closer through the growing stock and flow of GHG emissions.

Our essential collective challenge is to recognize the climate change crisis for what it is: The single greatest threat facing humanity. And the threat is growing ever more urgent. '[I]f we do not get this right, nothing else matters' (Mackintosh, 2019).

Bending the curve of climate change GHG emissions and pursuing net-zero carbon emissions by 2050 is, in 2021, the stated global diplomatic goal, agreed by most states and their leaders. Yet the goal is still in doubt, stalling efforts to take the necessary steps. To get from the polluting murky present to a sustainable future, to ensure a just transition and create a new Green Globalization 2.0, powered renewably and shared more equitably, is a truly mammoth task. It is not, however, impossible. To achieve net zero, we need to make war on carbon; redirect government policies and practices to net-zero goals; set guardrails and the glidepath to net zero; align all our business practices, incentives, and penalties to the goal; and set short-, medium-, and long-term targets on this journey.

Markets and their operation must be harnessed to the task of decarbonization, for markets can speed the rate of green transition, amplify public policies, accelerate their effects, and help embed and reseed a greener economic narrative across sectors and industries. In this economic reimagining, a balance needs to be struck between market imperatives and planetary essentials, for without the latter the former cannot operate. There is a need for 'politics and economics to enter into a frank dialogue in the service of life' (Francis, 2015). Economics needs to serve the planet, shake off adherence to erroneous worship of neoliberal tropes, and recognize that economics and the economy must serve society's goals.

Crises once recognized can become crucibles of change. They alarm and might even terrify us, but they demand action and force us to respond, focus, problem solve, and seek new ways of addressing rising dangers. Crises create new options and possibilities and allow new political economy and business coalitions to form; they enlarge perspectives, shift viewpoints, alter alliances, and change markets, economies, practices, and personal and business strategies. Crises can be destructive, but they are also drivers of innovation, construction, reform, reimagination, and rebirth.

Covid-19 lessons for our climate change responses

While there are many aspects of the collective response to the Covid-19 pandemic that could be improved, there are, nonetheless, illuminating lessons we can apply to craft climate crisis economics solutions.

I identify the following 15 key lessons from the pandemic that are relevant to our climate crisis responses. All lessons can be applied to aspects of our collective societal, governmental, economic, and individual reactions to and actions on climate change.

Leadership in crises is always crucial

Leadership always matters. Sound leadership in crises is essential to successful policy response. The pandemic has afflicted all of humanity. No one has been spared entirely from its societal, health, and economic effects. Nonetheless, some countries have been struck harder than others, in part because of a failure of consistent, credible, trusted leadership. For instance, the US lacked both presidential leadership and a coordinated national response. Many tens of thousands of lives have been lost in the US because President Donald Trump failed to follow scientific advice, failed to require masks, failed to coordinate responses, and failed to track and trace. UK Prime Minister Boris Johnson at the outset displayed a similar lack of leadership and constancy on pandemic messaging, policy responses, and implementation. The outcome in the UK, as in the US, has been disastrous from both a mortality and an economic standpoint. Lacking consistent, trusted, compelling leadership in a crisis, both countries dealt with the pandemic in worse ways than many other advanced-economy nations (such as Germany, Japan, the Republic of Korea, and Singapore). The US and UK also underperformed compared to many lower-income countries (such as Thailand and Vietnam). The pandemic has taught us what we should already have known: Leadership always matters to policy outcomes.

In fact, in confronting climate change, strong leadership is the *sine qua non*. With leadership comes clarity on national goals, consistency in approach, appropriate policies, faster implementation, and enhanced credibility and predictability that can amplify policy reach and its impact. Trusted leadership accelerates action and pushes societies and economies towards net-zero goals. This is true for leadership at all levels – governmental, regional, local, or firm – if we are to achieve both climate change net-zero goals and GHG emission reductions.

Coordination is crucial to collective response

The pandemic has demonstrated the importance of national and international coordination and cooperation. States that had strong, well-understood national responses (Australia, Japan, Korea, New Zealand, Singapore) fared better and had far fewer infections and deaths. Coordination was lacking at the international level. The US abrogation of leadership slowed and stymied the international response. President Trump's hostility to international diplomacy and his withdrawal of the US from the World Health Organization were complicating factors in the pandemic response. Thus, a truly global crisis was met by a disorganized response. This contrasted markedly with the 2008 G20 response to the global financial crisis, when President George W. Bush and Prime Minister Gordon Brown secured agreement among the G20 nations on a massive, coordinated response, increased resources, and the necessary reforms.

The lessons for our climate crisis response and reform endeavours are clear. Global leadership and collaborative engagement are crucial to achieve global climate change goals. Specifically, American and Chinese leadership and cooperation are necessary to secure progress because of the size and scale of their GHG emissions. Absent these giant state actors, the collective climate change goals become almost impossible to reach. In contrast, with constructive engagement, new stretch commitments, and a new greening policy equilibrium, a climate change glidepath to net zero becomes visible.

Delay is costly – act now

The pandemic has graphically illustrated that delay when facing a crisis is especially costly and damaging. It is far better to act early and fast than to procrastinate and hope for the best. Donald Trump and Boris Johnson delayed responses to the pandemic, stalled closures, and sent mixed and confusing signals. A failure to act early, clearly, and consistently cost lives, confused the public, and led to much worse outcomes. In contrast, leaders who acted quickly and decisively were able to bend the curve, save lives, and mitigate the worst effects of the first and second surges, and speed an exit from the related lockdowns, and lessen the economic effects. We can see that, as Balmford *et al.* (2020) note, there is 'no substitute for early action' (p. 696).

The lesson for climate change actors and responses is this: Don't wait. Act now to accelerate the achievement of stretch commitments to net zero. Stop discounting the future. Evidence repeatedly shows it is much less costly to prevent than to cure the effects of climate change (Manzanedo and Manning, 2020). We need to take mitigation actions today, adjust stretch goals and tighten timelines, and implement policies consistently and clearly. We can see, after 30-plus years of procrastination, fossil fuel lobbying, political timidity, and broken promises, that these repeated delays on climate policy responses are very, very costly. The delays have steepened the glidepath to net zero, increasing the danger and severity of severe weather events, and pulled nonlinear climate tipping points towards us in time. We need to recognize the climate crisis.

In crises, what was once impossible becomes possible

Crises such as the Covid-19 pandemic demonstrate the ways in which options are altered, enlarged, and shifted when everyone recognizes there is a common emergency. What perhaps only days before was thought impossible becomes possible in a crisis, including the massive use of state power, a triggering of vast fiscal and central bank resources, and even the shutdown of the global economy. By the spring of 2021, advanced countries alone had spent a staggering US$14 trillion in emergency fiscal relief to keep their economies on life support and their populations safe, working at home or furloughed. Yet, in early January 2020, the notion that such vast sums could be committed across the globe, with support across the political spectrum, would have been considered preposterous. Crises, once recognized and understood, change the political economy calculus fundamentally. The fiscal and governmental pandemic response shows the scale of what can be done in extreme circumstances. When survival requires actions that are truly enormous, possible responses become magnitudes greater than anything previously imagined.

The lesson from climate policy response is that vast public resources, dramatic actions, and state power and authority can be brought to bear when the crisis is recognized and immediate action is seen as essential. If we can achieve such leaps and shifts for the pandemic response, we ought to be able to make the necessary leaps and shifts to address the climate change crisis, whose magnitude is even greater and more threatening than that of the coronavirus.

Climate change responses may appear all but impossible right up until the crisis is recognized. Then the leap is made, and a new political economy green equilibrium is within reach. This is an answer to those who say climate change action is impossible or that it remains too glacial and too insufficient. Push towards a crisis recognition across communities and actors. Suddenly a narrative tipping point can be reached that alters the dynamics and rate of response dramatically.

Indeed, I believe that in 2021 we are on the cusp of a possible climate change narrative and policy tipping point. The exact point of final inflection is difficult to judge, but we know from past crises that when the narrative tipping point occurs, it is suddenly upon us, surprising and dramatic. For instance, the 2021 change in administration in the US and the US and Chinese net-zero commitments are crucial contributory, and possibly pivotal, shifts in the diplomatic and policy consensus, helping to move the climate change policy narrative from one destructive denialist state to another dynamic, action-focused state.

Governments are our last-resort actors when disaster strikes

The Covid-19 crisis demonstrates that when the grim reaper comes knocking, there are relatively few libertarians to be seen. When death stalks the land, most of us look to and demand government action to address the crisis and save us. Governments, in the advanced countries, responded by spending in excess of US$14 trillion in fiscal support and our central banks provided almost unlimited liquidity to markets. There was indeed vocal opposition to medical and economic lockdowns. However, the extent to which governments have been relied upon and have acted, with the measures being supported and observed by most citizens, are the more striking phenomenon, rather than the resistance in some countries and among some groups.

Crises clarify and underscore the central, ongoing pivotal role that governments play in securing our commonwealth, health, and welfare. Markets cannot and will not save you when all the chips are down; they seize up and fail when crises loom. Only the government, as your agent and servant, can marshal the necessary resources, power, and authority to navigate a route to health, economic or ecological safety, and stability.

The climate crisis echoes and repeats this essential lesson. We need and must rely on governments to help us achieve our common goals and secure our planetary and climatic safety and resilient economic security. Climate crisis response rests fundamentally first upon government action, regulation, oversight, the setting of guardrails, glidepaths, and agreed collective goals. The pandemic also reminds us that markets rely on government action, direction, regulation, and oversight to operate in support of our common societal, health, economic, and ecological goals. In securing our common climate goals, we need a clear-eyed realization of the continued crucial role that governments must play in setting the stage for collective response to crises and recovery, economic and ecological.

Institutions and trusted experts matter in crisis policy responses

The pandemic has showed us all the importance of trusted institutions, international and national. These organizations inform the public, advise policymakers, and help implement policy decisions. The World Health Organization was an essential player, albeit maligned by US policymakers. The EU performed an important coordination role, not only for the members states but also as the host of the key international conference that raised funds for global development and deployment of a vaccine. At the national level, trusted institutions and their leading experts, such as the US Centers for Disease Control and Prevention and Dr Anthony Fauci, director of the National Institute of Allergy and Infectious Diseases, were essential conduits for unbiased, trusted information. In a crisis, the public wants to hear the facts from trusted intermediaries. They don't want spin and obfuscation.

The lesson for climate policy design and action is that institutional constructs matter, below the level of the Conference of the Parties (COP), in the communication of policy goals and their consistent implementation. The climate crisis must not be a political football. The facts are clear and must be made understandable. The scientific facts and policy goals must be communicated clearly and consistently by institutions and experts that the public can trust and rely upon. Existing institutions will be used, strengthened, and repurposed to achieve climate goals. As necessary, entirely new constructs may also be useful and should be created. As we grapple with the transition to net zero, we should not assume the architecture will remain the same. Institutional innovation may be needed to help ensure our net-zero goals are being implemented, adjusted, and complied with across countries, markets, and economies.

A great weight can be borne by all once a crisis is understood

The pandemic has illuminated the remarkable fortitude of common people. It has showed the willingness and ability of the average citizen to bear great burdens over a long period, often without complaint, according to the needs of their societies and communities. Entire cities, regions, and countries were put on lockdown. Activities were restricted, travel halted, and meetings even with elderly loved ones prohibited. All work stopped or was made distance.

The willingness of billions of humans to collectively bear such a heavy burden is a remarkable testament to humanity's social, ethical, and moral nature. Many of us, certainly a majority, bore burdens for people we did not know and who we will never meet. There was some vocal opposition to the health restrictions, but most carried a weight so fewer would die, by observing lockdown restrictions, by changing their habits, by working remotely, by limiting their contacts with family and others, and much else besides.

Both the pandemic and climate change crises require, as Balmford *et al.* (2020) state, 'decisionmakers and citizens to act in the interests of society as a whole and in the interest of future generations' (p. 970). This lesson – that most people are cooperative and altruistic when a severe, recognized crisis strikes – is instructive and positive. Communities and individuals are capable of much more than is normally assumed in non-crisis times. We have seen this selflessness in other periods of war and privation. Societies and individuals can together bear great weights when their collective survival is understood to be at risk.

Individual responsibility is required to achieve our goals

The pandemic response relied on governmental action, direction, and resources, but our collective health goals could not be achieved without a great deal of sustained personal responsibility for actions and inaction, across societies. Countries with a strong sense of personal responsibility balanced by a strong sense of societal cohesiveness and collective responsibility fared better. Japan had no lockdown or work stoppage and has the lowest death rate in the G20. As of January 2021, less than 2,000 Japanese people had succumbed to the pandemic in a year. Japan achieved its remarkable public health outcome because the Japanese people followed health advice (masking, washing hands, self-isolation). Other countries suffered exponentially higher death rates in part because of weaker societal bonds and lower adherence to, or rejection of, health requirements and practices. Thus, American individualism and irresponsibility resulted in nearly half a million deaths (as of mid-January 2021).

This pandemic has demonstrated that our climate change goals can only be achieved if the net-zero goals and glidepath are supported and underpinned by the individual actions of citizens across the globe. The public needs to be able to visualize the science and to better understand how to change individual actions in response to climate change risk (Baskin, 2020). With a clearer understanding comes greater responsibility and shifts in what we do, buy, and eat and how we act towards one another, the planet, and our communities. Ultimately, however, we all have a personal responsibility to act and to internalize climate change risk and in doing so help us all achieve net zero and a liveable Earth for our children and grandchildren.

Fairness is essential and demanded

The pandemic underscored the importance of fairness in our interactions and, thus, in burden sharing. When crises strike, we can carry a greater burden, but people demand that all do so equitably, that there be little or no special treatment, or one rule for the rich and another for the poor. Recall the furore in the UK when Dominic Cummings, special advisor to Prime Minister Johnson, broke the lockdown laws to suit his own personal life. Public anger was widespread and visceral. People require fairness in burden sharing, not special treatment for some.

The climate crisis also requires us to recognize this inbuilt evolutionary fairness and social requirement in climate change policy, nationally and internationally. For only by ensuring fairness and a just transition can we all manage a smoother flight towards net zero. Fairness and a just transition are not an afterthought; they are an important structural element in the construction of a viable route to net zero.

We need to find ways to communicate a sense of urgency to all

The pandemic illuminated that it was important for citizens to understand the danger the virus posed. This can be particularly difficult early in the spread of a disease. Yet this is precisely when across-the-board action is needed to halt infections at minimal cost. This communication is difficult because people find it hard to understand nonlinear exponential growth. Moreover, citizens in some countries (such as the US) did not have a frame of reference to draw upon and found it difficult to gauge risk. This resulted in an underestimation of the risks posed by the virus among some populations. This underestimation, poor conduct in terms of wearing masks and keeping social distance, and lack of risk avoidance combined to increase the eventual infection and mortality rates. Other countries performed much better, especially those with direct experience of severe acute respiratory syndrome (SARS) (such as Japan, Korea, and Singapore).

The lesson here for climate change policymakers is that we need to devise new mechanisms to communicate climate risks and costs. As Pope Francis (2015) stated, we need 'a new dialogue about how we are shaping the future of our planet. We need a conversation which includes everyone, since the environmental challenge we are undergoing, and its human roots, concern and affect us all' (p. 14). The climate change crisis needs to be explained in ways people can understand, internalize, consider, and respond to. These narratives and stories can help crystallize the scale of the threat and move people to act. We have examples of best practice we can follow. Communities need to start 'the iterative process of narrative forming thorough a constructive dialogue ... this process should engage as many stakeholders as possible' (Bushnell, Workman, and Colley, 2016: 1).

Climate change policymakers and actors may need to create new forums and methods of dialogue that depoliticize the climate change narrative and stories we tell one another. We need to ground the discussion on understood and agreed and illustrated facts, which properly stated are indisputable. Once the facts are agreed, the urgency for action becomes self-evident and the discussion can then shift from accusation and counterclaim to a dialogue that centres on what needs to be done and who should do it, not whether a climate crisis exists at all.

Facts matter and must be defended and reiterated

The fake news pandemic that spread across the internet in parallel with the real Covid-19 pandemic was sobering and alarming. Tens of millions of Americans and hundreds of millions elsewhere across the globe consumed misinformation fed to them by artificial intelligence (AI) algorithms, sending them down ever darker rabbit holes of fake news and distorted clickbait views. This led people to believe masks were ineffective or counterproductive, to doubt the existence of the pandemic, and to question the safety and efficacy of the vaccines. A significant section of the public in many countries appeared to be receiving their 'news' from highly problematic sources such as Facebook, YouTube, Twitter, and Instagram. This pollution of the information atmosphere made communication more challenging and difficult.

The communication lesson for climate change action is that policymakers and actors need to underscore and reiterate the facts about climate change in as many different forums as possible, in clear and unpolitical terms. This fact-based dialogue must continue even if the evidence is (to scientific eyes) already indisputable and settled. There is a need for constant reiteration and discussion, across our societies, throughout our communities, societies, and economies.

Crises, once recognized, can vastly accelerate the rate of societal evolution

The pandemic has accelerated many dynamic societal and economic trends already underway. The accelerated shift to internet retail has been turbocharged. The use of videoconferences for meetings has become standard. The move to remote working has gone from the occasional to the norm. Now that many service industry chief executive officers (CEOs) see the enhanced productivity of their workers (and the lower cost of real estate), the shift to remote work may become a permanent shift for many. Or consider the collapse in business air travel, still in 2021 at less than 10 percent of what it was in late 2019. Few airline executives expect a return to business as usual in 2021. People will still fly, but not as often, cutting GHG aircraft emissions. The pandemic and economic crises have accelerated many such dynamics. No one expects a return in 2021 to the status quo ante.

A similar accelerative dynamic is beginning to become visible and must be supported in the climate crisis response and its impact on our economies and ways of doing business. Speeding the rate of this shift is essential, possible, and urgent.

Markets, once galvanized, can act and act fast

At the outset of the pandemic, it was commonly remarked that the average time taken to design and roll out a vaccine for a virus could be up to ten years, and that it would take many years, at a minimum, if we could get a vaccine at all. Yet, pharmaceutical companies across the world, pushed, encouraged, and supported by huge government contracts, raced to find a vaccine. Remarkably, by January 2021, seven vaccines had already been created, were in production, and were being injected across the globe. This is a truly astounding achievement demonstrative of what private firms and public authorities can achieve when the collective and private goal is clear and the need is extremely urgent.

This is an important lesson for climate change market dynamics. Once the broad climate change regulatory goals are set and are clear, predictable, and credible, individual firms can pursue them swiftly, seize the economic opportunity, and in doing so pull forward GHG reductions. The pandemic has shown us that, given the right incentives, clear goals, and guardrails, the private sector can be a climate change net-zero accelerant. We should conclude that 'climate change is not inherently in conflict with economic growth' (Baskin, 2020). Rather, our economies, investors, and actors, when incentivized effectively, regulated, and overseen, can help us achieve climate GHG goals.

Firm-level commitment and engagement matter

The pandemic demonstrated that markets and firms are essential actors in achieving the health goals set by the government. Most firms understand they operate with a social and economic licence and must act according to regulations and societal expectations. The pandemic changed business operation requirements and rules. Some firms closed. Many more continued to operate in the new pandemic normal, shifting strategies, products, approaches, and practices.

The lesson for climate change from the remarkable flexibility and changes that businesses across the world have initiated to handle and survive the pandemic is clear: Businesses can make significant changes if they are properly and clearly communicated, understood, planned, and executed. In fact, the shifts that are required to achieve net-zero goals are relatively long-range, though revolutionary, and are achievable when spread across a three-decade or longer time frame.

Crises force a reappraisal of what we value

The final lesson concerns the question of value. The pandemic has forced us all to consider what has value, what is essential, and what is not. We have recognized the value of workers we hitherto took for granted: grocery workers, delivery workers, eldercare workers, healthcare workers, and emergency medical technicians. Many of these groups are underpaid and had been underappreciated. The pandemic forced us to reconsider and recognize the worth of others. Through the pandemic lens, a grocery worker is far more important than an economist, and rightly so. I must have food to live. The quarterly economic forecast can wait.

Drawing on this value reappraisal from the pandemic, the climate crisis should push us towards a similar epiphany. We need to reassess the value – both monetary and non-monetary – of our economies' resources and peoples. We also need to reappraise how we value economically, emotionally, morally, and psychologically our land, air, water, and oceans. We need an enlarged ethical and moral calculus when we consider the value of others, human and nonhuman.

The structure of the book

In this book, these 15 crisis lessons will reverberate. There will be many instances explored where these lessons are being learned and are being reflected in actions by leaders, governments, communities, economies, markets, firms, and individuals. We will also see many instances where a great deal still needs to be done. We must recognize the climate situation for the crisis that it is. We can then begin to apply climate crisis economics and political economy solutions to the myriad of challenges we face on the glidepath to net zero.

A word about climate crisis economics. By climate crisis economics I mean a mix of economic and political economy policies and solutions needed to address climate change and accelerate the urgent and overdue transition to net zero.

Applying climate crisis economics solutions over the short, medium, and long term is essential if we are to hasten the transition and alter societal narratives and economics in ways that are necessary to avoid a hothouse world.

Practising climate crisis economics requires us to be realistic about our economic narratives and to revisit and reconsider our slavish adherence to neoclassical concepts that do not fit how markets operate and perform. Being frank and realistic about the limits of free markets and the need for oversight and regulation, especially when dealing with climate change challenges, leads us to different conclusions. These conclusions are not at odds with the operation of free markets but, rather, require an adjustment of the parameters of decision making and of our assumptions about the incentives and penalties.

Delivering effective climate crisis economics solutions requires us to use models that answer the key question before us: 'How can pricing signals and models help smooth the path to the net-zero goal leaders and governments have agreed is essential?' Modelling climate change is fraught with forecasting difficulties, political assumptions, and value judgements. To be effective climate crisis economics advisors, the models we rely on should be scrutinized and judged based on whether they help policymakers achieve net-zero transition goals.

If the dominant model produces temperature outcomes way outside what is needed for human and planetary survival and sustainability, we need to revisit the models and reconsider our assumptions. If the model fails to account for real risks and political and economic dynamics that adversely affect climate outcomes, we need to ask whether we need to design new models or use new assumptions. Chapter 2 addresses this matter of cloudy horizons, middling models, and problematic assumptions and asks whether the dominant model is fit for purpose as we seek to secure net-zero climate goals and transform the economy.

Chapter 3 addresses signs of more ambitious net-zero climate goals from leading powers. Taken together, they may be turning what appeared impossible into possible and reachable, albeit a difficult and still distant net-zero goal. Leadership and ambition matter. Here, the news, although challenging, is not all bad.

Ensuring we bend the curve on GHG emissions means climate crisis economics must tackle head-on the greatest failure of markets vis-à-vis climate change: The market failure to internalize the cost of carbon at a rate that reflects the real, current, and rising damage to the planet. Climate crisis economics will fail if we do not wage war on carbon and price carbon progressively and in a manner that shifts incentives, expectations, and market reactions and strategies. Chapter 3 lays out the carbon pricing options and presents cases where such policy solutions are already being applied. It assesses taxation levels, emissions trading schemes, and evidence that we can bend the curve of GHG emissions, pulling forward and accelerating the rate of change. Carbon pricing alone is not enough to deliver the climate crisis economics solutions we need. We also must foster a change in the stories that market participants tell one another. Pricing can help to start that process, but it is not the only positive feedback loop.

Market incentives and dynamics must also shift and evolve rapidly to adjust firm-level strategies in the real economy and markets. Here, the shift is already underway and gathering speed. Demographic and societal dynamics are acting in support of our climate change goals, rewarding first movers, punishing laggards, shifting returns, and altering expectations and outcomes. Chapter 4 highlights evidence of positive market dynamics that are beginning to support the rate of transition and are pulling forward GHG emission reductions.

Setting ambitious net-zero stretch goals is the starting point for a narrative and policy shift, but ultimately consistent implementation matters most of all. Chapter 5 sketches the outlines of institutional innovations that can help ensure goals turn into facts on the ground, shifting from exhortations into credible, predictable policies, well understood by markets and individuals, that are effectively overseen and monitored.

Many of the technologies and mechanisms needed for climate crisis economics and political economy solutions to deliver already exist. Unfortunately, most are not being implemented and applied fast enough. This must change, supported by a greening of national industrial policies. Governments will not pick individual winners and losers, but they can and should progressively green the regulations, incentives, and frameworks within which markets operate. In doing so, governments can redirect the animal spirits of markets to the benefit of the planet. Governments have a key role in supporting innovation and the rate of new technology diffusion, in steepening the S-curve of adoption so that, sector by sector, market by market, old polluting practices are replaced by carbon-neutral and negative circular economic models. This is a massive task, with many sectors having only barely begun their net-zero journeys. Here again, well-designed regulations and incentives are crucial to set the glidepath and mile markers and to spur the rate of diffusion and decarbonization. Chapter 6 focuses on the challenge of diffusion, identifying examples of leaders and laggards.

As countries set stretch goals, speed the transition, implement policies, and design institutions to assist and spur technology diffusion, it will also be crucial to reimagine our climate change narratives and stories, to extend the dialogue and discussion. We need to find ways to talk to one another about the risks of climate change, agree on the facts, and turn to the solutions. We have examples of how to do this. In many locations, this renewing of our climate change narratives is underway. In other, more fractured societies, a great deal remains to be done to repair understanding and support dialogue. Cities and localities across the globe have shown how it can work. Young people have taken up the challenge and engaged in the net-zero debate and struggle. Elements of the new narratives are audible and need to be amplified. Chapter 7 discusses the centrality of narratives to the climate change net-zero journey.

The future need not be dystopian. We can construct a Green Globalization 2.0. This transformational economic shift has already begun. Green Globalization 2.0 will be the engine of a greener, more prosperous, more equitable future. There is no inherent conflict between economic growth and securing net zero. To the contrary, the former is essential to achieving the latter. Sustainable, resilient, broad-based growth will accelerate towards Green Globalization 2.0. Green Globalization 2.0 can be grounded on and include consideration of not only the monetary value of that which we price but also the moral and ethical value of our planetary landscapes and ecosystems. Chapter 8 lays out the emerging contours of that potentially more moral and ethical, sustainable Green Globalization 2.0.

In *Climate Crisis Economics*, I argue that once a tipping-point plurality of communities (geographic, political, economic, and societal) recognize the climate crisis for what it is – urgent, immediate, and the singular challenge of our lifetimes – we can act to address it and secure the route to net zero and create Green Globalization 2.0. Crises are crucibles of change. Once recognized and understood, they trigger swift, cascading shifts in policy action and of economic, business, and personal decisions. Suddenly, addressing the climate crisis becomes achievable, as all collectively begin put their minds and labour to the task.

This book is not a story of failure. Rather, it shows where we are succeeding, who is leading, what needs to be done, what is already underway, and, yes, the many areas where progress is too slow. *Climate Crisis Economics* argues that we can get to a decarbonized net-zero future by 2050. I believe the green globalized economy of the future can be better and more equitable, sustainable, resilient, and dynamic. But this transition cannot happen spontaneously. All of us – as governments, markets, businesses, and individuals – must align ourselves to the net-zero goal. We all have a role to play.

Note

1   The term Anthropocene epoch was invented, as its creator, the late Nobel-winning chemist Paul J. Crutzen, said, 'on the spur of the moment', to describe a new geological epoch in which 'humans were having such a profound impact on the planet that it was time to recognize a new geological epoch – the Anthropocene' (www.washingtonpost.com/local/obituaries/paul-crutzen-dead/2021/01/29/97e9c200-6244-11eb-afbe-9a11a127d146_story.html).

References

Balmford, A., Fisher, B., Mace, G.M., Wilcove, D.S., and Balmford, B. (2020) 'Analogies and lessons from COVID-19COVID-19 for tackling the extinction and climate crisis'. *Current Biology*, 30 (7 September): 936–983.

Baskin, K. (2020) 'Four lessons from COVID-19 to help fight climate change'. MIT Sloan School of Management, 22 June [Online]. Available at: https://mitsloan.mit.edu/ideas-made-to-matter/4-lessons-COVID-19-to-help-fight-climate-change (accessed: 22 January 2021).

Bushnell, S., Workman, M., and Colley, T. (2016) 'Towards a unifying narrative on climate change'. Grantham Institute Briefing Paper No. 18, Grantham Institute, Imperial College of London, London.

Carney, M. (2020) BBC Reith Lectures. Lecture 4 [Online]. Available at: www.bbc.co.uk/programmes/articles/43GjCh72bxWVSqSB84ZDJw0/reith-lectures-2020-how-we-get-what-we-value (accessed: 14 January 2020).

CBS. (2020) 'The record-shattering 2020 hurricane season, explained', 20 November [Online]. Available at: www.cbsnews.com/news/atlantic-hurricane-season-2020-record-breaking (accessed: 19 January 2021).

Francis. (2015) 'Laudato Si' [Online]. Available at: www.vatican.va/content/francesco/en/encyclicals/documents/papa-francesco_20150524_enciclica-laudato-si.html (accessed: 14 January 2021).

G30 (Group of Thirty). (2020) *Mainstreaming the Transition to a Net-Zero Economy*. Washington, DC: Group of Thirty.

Littlemore, R. (2009) 'A review of Michael Mann's exoneration'. *DESMOG*, 4 December [Online]. Available at: www.desmogblog.com/review-michael-manns-exoneration (accessed: 19 January 2020).

Mackintosh, S. (2019) Discussion with a leading economist. Author's notes, December.

Manzanedo, R.D. and Manning, P. (2020) 'COVID-19: Lessons for the climate change emergency'. Science of the Total Environment, 742: 1463–1465. (10 November). [Online]. Available at: https://doi.org/10.1016/j.scitotenv.2020.140563 (accessed: 17 May 2021).

Scientific American. (2020) 'NASA says 2020 tied for hottest year on record', 14 January [Online]. Available at: www.scientificamerican.com/article/2020-will-rival-2016-for-hottest-year-on-record (accessed: 19 January 2021).

Thunberg, G. (2019) Speech at the UN Climate Action Summit, New York City, 23 September. Transcript [Online]. Available at: www.usatoday.com/story/news/2019/09/23/greta-thunberg-tells-un-summit-youth-not-forgive-climate-inaction/2421335001 (accessed: 14 January 2021).

2

OBSCURED HORIZONS AND MIDDLING MODELS

Climate crisis economics requires us to look towards distant horizons to try to grapple with an unprecedented global problem of gargantuan scope. It requires humanity to comprehend an almost infinite number of interrelated factors that are at once all-enveloping and often subtle, as we try to understand a process that is simultaneously as fast as a raging forest fire and as slow as a sequoia's growth. This is difficult to do. Our natural tendency as individuals is to think of ourselves, our family, our community today, and perhaps tomorrow, but 10 or 20 or 30 years from now? No so much.

Extend our horizons

Yet dealing with the climate crisis requires us to extend our economic and societal horizons beyond the individual, often ill-informed short-term decisions to encompass consideration of the common good and to extend our empathetic horizons to communities and families far removed from our own experience in distance, culture, and practice.

We can make this leap when emotional ties to each other and the Earth are strong, when we understand one another's struggles, and when leaders help us tap into 'the better angels of our nature' (Pinker, 2012) rather than the devils of nationalism and nativism.

Effectively addressing climate change requires societies and individuals to expand their horizons and their understanding of the unfolding crisis, to enlarge their notion of the collective good, utility, and burden sharing, and of the need for action at every level – local, national, and international – from their family to their community, to the global community. We need to understand that we are all in this together and we all have a collective responsibility to each other and to Planet Earth, and we need to work together on our journey to net zero.

DOI: 10.4324/9781003037088-3

Understand the limits of our models

Our economic models and narratives must be fit for purpose if they are to help us achieve the climate change goals that our societies have committed to. Too often, our economic models are not up to the task. Whether due to reasons of design, faulty assumptions, ideological bent, narrowness, or simply the limits of forecasting on a subject as complex as climate change, our models can undermine essential market, policy, and personal shifts, and stymie action. In short, some of our macro models are out of step with what governments and increasing numbers of businesses, voters, and our children are demanding. Better models are needed to illuminate the path forward.

Is a model supporting the policy goal?

We must ask ourselves: 'Is this model helping us achieve the net-zero goal which scientists tell us is essential to our continued survival?' If a model's assumptions lead to results that are far outside the temperature goals that the governments of the world are seeking – i.e. the Paris Agreement goal of limiting temperature rises to 1.5 degrees Celsius above that of preindustrial times – and instead produces a so-called 'optimum' result that forecasts a rise of 3 to 3.5 degrees Celsius or more, we must ask: 'Is this really an optimum planetary outcome?' The answer is, no, it is not.

Climate crisis economics must also be focused on helping to design routes, levers, and mechanisms to speed the glidepath to net zero and achieve as close to the Paris Agreement temperature goal as possible. If a model or its assumptions do not help us get there, we need to set it aside and change our inputs and approaches.

This chapter argues that the dominant model used to forecast the economic impact of climate change has undermined our climate goals and contributed to the failure of policymakers to respond with sufficient urgency to the climate crisis that draws closer every day.

Worry more about fat tail risks

The way we think about and use our models must help us achieve climate change policy goals as well as inform us of the price of doing so and of failing to act. The planetary stakes could not be higher, and we must be alert to fat tails – i.e. scenarios with relatively high probability of extreme outcomes and alarming nonlinear tipping points and breaks that could move the planet's climate from the current fragile equilibrium to a hothouse future. If our models cannot adjust to include such dynamic systems, then we need to look for new constructs and new modes of thinking about the climate crisis, including being realistic about what it is and is not possible to do with economic modelling on a planetary scale.

Integrate economics with ethical and moral judgements

Addressing climate change requires us to fuse economics with scientific data and planetary warnings. We need to consider issues of morality, ethics, and steward-ship, not only short-term personal utility, self-interest, or self-destructive behaviours that do not factor in societal considerations and obligations. As Pope Francis has stressed, our economies must serve our societies and ensure our planetary survival. Doing so requires us to make moral and ethical decisions as well as economic ones. Economies serve society and should not be impervious to critique, regulation, or redirection.

The first challenge in addressing climate change is to extend our horizons; that is, we need to move from our obsession with tomorrow or the next economic growth figures or the next week's unemployment figures to the long term. In doing so, we begin to gain a better perspective on the problem, its scale, and what should be done about it. Lifting our gaze to a distant horizon is not easy, for we have no framework or examples to draw upon.

A tragedy of horizons

It is extremely difficult for human beings and societies to comprehend the global scale and long timeline of the unfolding climate crisis. We suffer from what Mark Carney, former governor of the Bank of England and now the UN Special Envoy on Climate Action and Finance, calls a tragedy of our horizons. Carney notes:

> We don't need an army of actuaries to tell us that the catastrophic impacts of climate change will be felt beyond the traditional horizons of most actors – imposing a cost on future generations that the current generation has no direct incentive to fix.
>
> *Carney, 2015: 2*

Most of us cannot see far enough ahead to understand the scale of the long-term climate change horrors that might affect the planet. The timescales are too long for our minds to grasp. We cannot say, 'I remember what we did in the last global warming period'.

No examples outside myth, archaeology, and ancient history

Outside of Old Testament fables of Noah's Ark, passed down over thousands of years, there are no recorded instances in human history of sudden, irrevocable, dis-astrous changes in the Earth's climate. Noah's flood is believed to have occurred when the Mediterranean Sea flooded the land that separated it from the Black Sea, flowing south to north across the isthmus. This flood may indeed have occurred, but we are 'uncertain and [the examples] lie outside the bounds of human experience'

(DeFries *et al.*, 2019: 5). Science tells us sudden climate shifts have occurred, but they are lost in the mists of time.

Palaeoarchaeology illuminates, for instance, the Younger Dryas period 11,000 years ago (NOAA, 2020), when sudden climate shifts occurred, demonstrating our planet's climate can change from one equilibrium to another well within a single lifespan. Rising temperatures in this period caused the North Sea to swallow up a highly fertile European landscape of mountains, rivers, and settlements on the Dogger Bank, a landmass the size of the Netherlands, a land bridge upon which my prehistoric hunter-gatherer ancestors could have walked from Edinburgh to Oslo, but which vanished below the North Sea.

Leaping millennia forward, archaeology and historiography suggest the collapse of the Roman Empire was in part linked to a period of warming, severe drought, and multiple years of failure of grain harvests in Egypt. Rome, whose massive population relied on grain imports from Egypt, olive oil, and wine from Spain, and much else besides, was destabilized. The warming climate drove climate refugees and tribes across the empire's outer borders to clash with and defeat Roman legions, further destabilizing the empire as it struggled to feed its population (Harper, 2017).

Of course, these mythical, Palaeolithic, archaeological, and early historical examples of the effect of climate change on human societies are not today part of our common narrative and understanding. In 2021, society has no climate parallels. Because of this, we have great difficulty grasping the enormity of a building global crisis caused by our individual and collective actions over the years and decades.

Yet securing a net-zero future requires societies, governments, and individuals to extend their political, societal, economic, and personal horizons. We need to commit today to goals whose costs will be borne by us but whose results will only be fully visible decades from now. So while we must be made aware of the costs and burdens of achieving our climate goals, we also need to make achieving our short-, medium-, and long-term goals manageable and their benefits visible. Governments play a key role here. They can help stretch our horizons, plan long term, and help us address the 'tragedy of the commons', which is defined as

> a problem in economics that occurs when individuals neglect the well-being of society in the pursuit of personal gain. This leads to over-consumption and ultimately depletion of the common resource, to everybody's detriment.
> *Boyle, 2020*

The tragedy of the commons confounds the climate change goals we must secure and is something the COP process has tried to address through coordinated implementation of GHG emissions targets. Governments can assist, as well, by helping to change our narratives on climate change response, transition, and a net-zero industrial regreening regrowth and renewal.

Think of a village common as pastureland on which villagers can graze their animals. Rational but self-interested individuals will want to accumulate wealth, so each places a head of cattle on the commons. Eventually, as each villager adds another head of cattle, the common's most important resource, grassland, becomes depleted and the pastureland denuded. Everyone loses. This is the tragedy of our planetary commons. In this case, 'ruin is a destination toward which all men rush, each pursuing his own best interests in a society that believes in the freedom of the commons. Freedom of the commons brings ruin to us all' (Hardin, 1968).

Climate change is the ultimate problem of blinkered horizons and a tragedy of the commons, the classic global public good, because each country's GHG emissions contribute cumulatively to the increase of the overall stock, but each country's abatements may potentially entail higher costs than immediate direct benefits unless effective, collective actions are taken to stop freeloaders from polluting now at the cost of other, usually poorer, more marginalized, less powerful countries and groups.

Understanding how limited our horizons are and trying to expand them by lifting our gaze is difficult, as is addressing the tragedy of the global commons. Both challenges require government, economic, business, and personal actions to correct. It is governments that set the guardrails, the glidepaths, and the rules for societies and economies. Governments acting in a collective and coordinated way are essential to successfully lengthen our horizons. That is what the COP process and negotiations are meant to evoke. So far, they have had middling success, but in 2021 there now seems to be a greater chance of success.

Economics, too, must shift and recognize the biases of its own narratives and the limitations of its models and ways of thinking to more effectively confront climate change and to help design policy options that work. In short, a rebalancing is required.

Economic stories inform our personal and policy actions

Economics has its own narrative strands. Shiller (2019) defines an economic narrative as 'A contagious story that has the potential to change how people make economic decisions' (p. 3). He identifies seven forms of economic narrative. These narratives grip us and move us – from tech boom to bust, from credit-fuelled great moderation to global financial crisis, from a bitcoin boom to bust and to boom again. We are often not entirely aware of the narrative as it pulls us along and alters our thinking and actions. Many of the narratives are short-cycle stories, such as 'Buy bitcoin now'. Some are longer cycle but still important, such as 'house prices can only go up'. Such stories compel individuals to actions that swell bubbles and cause collapses. Other economic stories run over many years and embed themselves within our policy dialogues and economic systems.

Economic ideologies have narrative waves and currents, from classical to Keynesian to neoclassical and to neo-Keynesian; from supply-siders to demand-siders. The relative strength of economic ideological narratives waxes and wanes. Classical economics dominated in the early part of the twentieth century, informing how governments responded to the Great Depression. President Hoover, for example, declined to use the state to avoid catastrophe, and his economic and ideological failures led to historic hardship in the United States.

US President Franklin Delano Roosevelt offered an activist state response and was swept to victory three times in the 1930s and 1940s. FDR brought the power of the state to bear in economics and planning and, with the build-up to war, reconstructed the economy after the depression. War economics was interventionist and directional – it was state industrial policy writ large – and this drove the Allies to victory. Economic markets assisted but came under strong oversight and direction of the wartime state.

Postwar economics and ideology supported a managed global system of fixed currencies and the gold standard. An international architecture, designed by John Maynard Keynes and his colleagues, was created by the US and its allies to reconstruct the global economy. Keynesian demand-side economics dominated the discourse throughout the 1950s and 1960s as baby boomers prospered, economies grew, and Europe and Japan were reconstructed with American government funds and support. Full employment was the goal, with government interventions and spending to secure sustain growth.

The 1960s gave way to the 1970s; stagflation took hold (in the UK), with economic stagnation and inflation occurring simultaneously. Countries struggled with social unrest, strikes, and other tensions. The US, mired in the Vietnam War, responded to its own economic malaise by abandoning the Bretton Woods system of fixed exchange rates. The UK faced terrorism in Northern Ireland, and a weak state was apparently unable to act to stabilize the economy or address social and economic ills.

As the economic woes mounted, neoliberal economists seized control of the narrative, with Milton Friedman, Friedrich Hayek, Keith Joseph, and others taking aim at a Keynesian narrative that appeared to have run its course. Instead, they championed the free market, a release of 'animal spirits',[1] supporting a decades-long deregulatory shift that altered the balance in economies in favour of markets. The state retreated.

The Austrian neoliberal ideological school won out in the US and the UK, and their market-centric narrative became firmly entrenched in the US-dominated international norm-setting institutions. This economic narrative shifted decisively away from state action, state support, and industrial policy and direction to a worshipping of unrestrained markets and market economics above other hybrids, although these hybrids survived (and sometimes thrived) in continental Europe and Asia.

From the late 1970s onwards, US President Ronald Reagan, UK Prime Minister Margaret Thatcher, and their ideological progeny (including centre-left leaders such as US President Bill Clinton and UK Prime Minister Tony Blair) held broadly to the neoliberal, market-centric mythologizing of the power of the invisible hand, the importance of market solutions, and avoidance of state activism in the economy and its direction. The markets knew best – always.

This dominant neoliberal narrative and its various facets comprise the economic story we still tell ourselves today and which economists tend to use when advising many leading governments in their policy process. However, this narrative limits the extent to which we can respond in guiding the path of the economy because it does not accept that urgent, collective planetary climate change action is needed.

Economists need to be frank about the limits and failures of their narratives and models and their failure to provide answers to the climate change crisis. We can then begin to apply solutions, be they economic (pricing), legal (antitrust and taxation), or regulatory (incentives and penalties), to the climate change crisis as we seek to construct a narrative model of climate crisis economics.

Economists and policymakers must reconsider the economic models used to understand the way forward on climate change. Unfortunately, the dominant model and its assumptions have undermined the speed of our collective response to climate change.

Climate change models – their limitations and uses

Economists have used Integrated Assessment Models (IAMs) to grapple with climate change for years. Early IAMs were energy economy models with a model of GHG emissions added to them. Today's IAMs, which are the standard tool to help visualize the effects of shifting climate change policy options and levers, include carbon cycle models and simplified climate models. Scores of economists use IAMs to try and gauge the impact of climate change and policies, but they remain controversial in their use and outputs. Pindyck (2017) argues that:

> IAMs have crucial flaws that make them close to useless as tools for policy analysis … [they] create a perception of knowledge and precision that is illusory and can fool policymakers into thinking that the forecasts the models generate have some scientific legitimacy.
>
> *p. 100*

Potential users need to be aware of weaknesses and limitations of IAMs. Climate change economic models are not reality. It might be better to adopt the position that 'all models are wrong, but some are useful' (Box, 1970). Potential users need to remember that a model's usefulness depends on the context in which it is used. If I use a flatwater canoe in whitewater, I will capsize. Similarly, if I use the wrong model for climate change or make erroneous assumptions in the model, the resulting output will be problematic and produce possibly incorrect conclusions resulting in dangerous policy decisions.

Models are at best simplified facsimiles of reality that we can manipulate to aid visualization of possible theoretical outcomes. These outcomes are fundamentally driven by the assumptions we choose (rate of growth, propensity to invest or save, current and future cost of carbon, and so on). The assumptions have economic and ethical implications. I may prefer to spend now rather than protect my children from harm, penalizing future generations. I might place no value on other species, denying the importance of nature and the ecosystems upon which we rely. Or I might instead value future generations equally, take a view of collective stewardship over the planet, and weigh costs and risks differently, thus altering my inputs into the IAMs.

Models are apolitical; our inputs are not

Models are apolitical, but our inputs are not. It is reasonable to debate political and ethical judgements. We all do it all the time in our work and decision making. In arguing the pros and cons of climate change mitigation and its costs and benefits, economists take implicit and explicit political positions. It is better to acknowledge the ethical and political judgements in the models we use. Model users should be clear about their political economy and ideological stances and admit they use models to help calibrate and achieve them. Policymakers should not believe using any particular model produces the sole optimum answer. Models are simply tools, and tools with many flaws. Models can aid decision making, but they can also stymie action if they deliver conclusions that slow necessary climate action and instead give fuel to the arguments of naysayers and denialists.

Rolling the DICE

The most well-known and influential IAM was designed by William Nordhaus in 1992. His Dynamic Integrated Climate-Economy model (DICE) won him the 2018 Nobel Prize in Economics and has provided the economic profession with decades of global estimates of the impact of climate change. It is the model most often used to estimate the costs of mitigation, the cost of carbon, optimal global temperature increases, and various glidepaths in the decades ahead. Its impact on the economic narrative on climate change has not been positive.

DICE is usable by nonspecialists and is free, downloadable, or usable online. It does not cover all the details of economic and climate processes. Rather, it combines a simple economic model with a simple climate model to provide a tool for students and policymakers to think about climate change policy and ways to use the levers they have available. DICE allows individuals to adjust assumptions on growth, savings, emissions, and so forth, and to see how the outcomes shift accordingly. It illuminates how policies can affect outcomes.

As an economic growth model, DICE assumes economic production occurs when labour and capital are available. Part of economic production is then invested to create capital for the future. The rest is consumed. DICE assumes that 'happiness' relies entirely on consumption today. That is, it assumes maximum benefit comes from spending. Just as with other models, the conclusions drawn from this model are only as good as the robustness and realism of the assumptions underlying it (Parramore, 2019). DICE, like all IAMs, is bedevilled by limitations related to the assumptions used as inputs. DICE is controversial because the outputs from some iterations of it have helped stall action, strengthened those who say climate change dangers are overblown, and undermined the drive to secure net zero. DICE's significant weaknesses include:

- The high discount rate problem
- The intergenerational inequity problem
- The growth assumption problem
- The damage function problem.

The following sections discuss DICE's weaknesses as the dominant climate model and economic narrative. Unfortunately, DICE has been partially responsible for 'the failure of the world's governments to pursue aggressive climate action over the past decades' (Hickel, 2016).

The discount rate problem

An essential first step in the climate crisis economics narrative is to recognize that we no longer have the luxury of discounting the future because the climate crisis future is almost upon us, and we have an ethical and moral obligation to preserve the planet and invest today to protect it for future generations. Discounting is used to translate future costs to current dollars. Discount rates put a present value on costs and benefits that will occur decades into the future. Discount rates are crucial for climate change policymaking because they help determine how much today's society should invest in limiting the impacts of climate change in the future. Historically, economists have preferred using the rate of that return markets provide for long-duration government bonds.

The discount rate is the key decision in DICE. It determines whether you spend, save, or invest now to mitigate future risks. If we use a high discount rate in the economic story we are constructing, the economic incentive to invest in climate change mitigation decreases. Nordhaus's 2007 DICE first used a very high discount rate of 5 percent. This led to the conclusion that spending for mitigation immediately was not worth it economically because the rate of return for saving, and thus inaction, was so high. DICE's policy conclusion – to wait and not bother mitigating climate change – may well have stalled policymakers' action on climate change. Subsequent model adjustments have lowered the discount rate to 3 percent. Even in its latest version, in 2018, the current DICE rate is more than double the rate recommended by Stern 15 years ago in his *Review on the Economics of Climate Change* (Stern, 2006).[2] Stern recommended a discount rate of below 1.4 percent. At this much lower figure, it makes a great deal more sense to remediate today to counter the future effects of climate change tomorrow because future dollars are not worth much more than current ones. Policymakers should have listened to Stern, not Nordhaus.

One should not give too much power to one model or one approach. However, we can hypothesize a DICE feedback loop. Early model iterations used a very high discount rate, suggesting little or no mitigation action was worth taking. This led to policy inaction informed by the model. That inaction allowed a worsening of GHG emissions and of the climate. Small model adjustments were made as the real climate deteriorated, but they were not enough to drive a meaningful policy shift.

BOX 2.1 PERNICIOUS USE OF DISCOUNTING – THE TRUMP CASE

Using a high discount rate deliberately skews a model's results because the discount rate applied makes a huge difference in the evaluation of policies (Fleurbaey and Zuber, 2013). Government planners understand this, and they understand that a high discount rate can contribute to slowing the momentum for policy action as it signals to all levels of government that investing now is not prudent and that waiting is the best option. President Trump, or at least some politically motivated, mathematically savvy officials in his administration, understood that inputs affect model outputs, and thus decisions.

This recognition was behind a 2018 revision in how the government calculates the cost of carbon. Trump officials first lowered the cost of carbon from an Obama estimate of US$50 per ton to US$1–US$7 per ton, making pollution pay more, and making mitigation seem not worthwhile. The Trump team then adjusted the discount rate upwards from (an already high) rate of between 2.5 and 5 percent to between 5 and 7 percent (*New York Times*, 2018). The Trump administration knew that using an excessively high discount rate in the cost–benefit analyses practically zeros out the rationale for preventive action because the model suggests you are better off saving the money than spending it now via investments to abate GHG emissions.

The Trump example demonstrates the distortive, ethically dubious, political damaging effects of high discounting.

Climate change policy inaction therefore is erroneously informed in part by the assumptions and outputs of the model. This is not a productive feedback cycle for economic modelling or the global climate story.

The Trump administration has vividly demonstrated how the misuse of carbon pricing and discounting can be deeply political and problematic (Box 2.1).

The intergenerational inequity problem: We value future generations – don't we?

The higher the discount rate, the less we value future generations' well-being, and the less we care about the future of the planet or future generations. This intergenerational approach by DICE amounts to discrimination by date of birth. It suggests the further away you are from me in time the less I care about you and your survival. This is ethically and morally questionable and not credible since we have finite resources and only one biosphere. It is also morally and ethically offensive as it penalizes future generations. Greta Thunberg's powerful castigation at the 2019 United Nations Climate Action Summit of the individuals, businesses, and leaders who delay real reform and discount her future with their inaction, rings in our ears:

You come to us young people for hope. How dare you. You have stolen my dreams and my childhood with your empty words … People are suffering. People are dying., Entire ecosystems are collapsing … We are in the beginning of a mass extinction …. For more than thirty years the science has been crystal clear. How dare you continue to look away.

Thunberg, 2019

DICE's high discounting rests on the faulty assumption that spending now is always preferred over investing today for tomorrow. Yet this does not match the way we behave in our own communities and families. Consider the vast amount of resources that families and communities commit to looking after and educating their children and grandchildren. In the US, a couple who gave birth to a child in 2015 will have spent US$284,000 by the time that individual graduates high school (USDA, 2020), and then another massive amount to send the child to college. Yet this cost is understood by families, who choose to invest in the future of their offspring instead of spending their money now or saving for a new kitchen, boat, or caravan or recreational vehicle. Families across the world go to huge lengths to provide education for, and safeguard the future of, their children and grandchildren. Most families (if they are able) prioritize such educational investment in their children's futures over gambling or drinking.

The climate crisis future is almost upon us. A child born in 2020 will have barely begun adulthood and a working life by the time our global carbon budget is exhausted. Just as parents, grandparents, aunts, uncles, and communities invest in our children, so too societies need to invest in climate mitigation and industrial transformation. By doing so, we can help ensure we are not poorer tomorrow. Conversely, if we do not invest, we increase the chance we will in fact be economically worse off in the future.

Will we all be wealthier tomorrow? The growth assumption problem

There is another flaw in DICE: The use of too rosy an assumption about the future rate of economic growth. Will we all be wealthier in the future? Perhaps not. Current rosy growth assumptions may be problematic, given the decades of secular stagnation that have afflicted the US, Japan, and Europe, including slow, subpar growth; very low inflation; weak consumption; and low productivity. Should we refuse to invest and spend today or save for an uncertain tomorrow? Is it credible to assume that decades from now the Earth and our economies will be relatively unaffected by climate change dynamics? This flies in the face of what we can observe as the climate warms. Might it be the case that failing to mitigate today, to invest in green technologies today, will instead increase the costs tomorrow, lower potential growth rates and worker and land productivity, and so on? For many countries and populations, it is more likely that tomorrow could be worse off than today, especially if we do not invest now to create the basis for net-zero industrial transformation of tomorrow.

Tomorrow we could all be poorer if we allow the current trends in GHG emissions and climate change dynamics to unfold unchecked. DICE's too-rosy assumptions and growth story ignore the increasing economic damage and disruption to hundreds of millions of people from climate change. These risks are already visible and the damage appears to be increasing, according to regular International Panel on Climate Change (IPCC) reports. IAMs such as DICE do not sufficiently include possible weakened growth caused by feedback loops from the climate crisis.

The rising cost of the economic catastrophes to come: The damage function problem

A crucial assumption in IAMs and DICE concerns the so-called damage function – i.e. the relationship between rising temperatures and growth and the extent to which GHG emissions flow and stock will hike temperatures and damage our economies. As with the discount rate, DICE model adjustments have gradually increased the damage function, as the world has failed to address GHG emissions, but does DICE capture the extent of the future damage? DICE looks at past costs of climate change and extrapolates forward from these, adding in a 25 percent 'fudge factor'.

DICE's economic damage conclusions are a recipe for policy inaction. Using his own damage function, Nordhaus (2017) concludes that:

> Including all factors, the final estimate is that the damages are 2.1 percent of global income at 3 degrees Celsius warming, and 8.5 percent of income at 6 degrees of warming.
>
> *p. 1519*

Climate denialists and naysayers rejoice. Lazy policymakers, too. Nordhaus is suggesting the economic cost is very low indeed, for a doubling of global temperatures above that recommended by the collective consensus of the global scientific and IPCC policy community. Nordhaus's higher damage estimate for a 6 degrees Celsius warming plays out over more than a century, and incredibly concludes that the economic cost is less than 0.1 percent GDP per year.

By setting the damage function so low, the implied recommendation to wait – to not invest now to mitigate climate change – is reinforced. After all the, damage is miniscule. As Alfred E. Neuman would say: 'What, me worry?' (Neuman, 1954).

The clear danger is that if we underestimate the damage, and we assume people will be wealthier in the future, the benefits from acting now again dissipate for policymakers and planners.

DICE's pernicious effect could, 'rather than providing sensible guide to policy, lull society into a false sense of security' (Keen, 2019). We should set aside DICE's economic impact conclusion and look at other estimates and evidence.

Evidence suggests many people, especially the poorest among us, already confront growing and life-threatening and economically costly impacts from climate change. The frequency and severity of climate-change-driven weather events are increasing. It is likely future weather events will get worse in years and decades, not centuries, if nothing is done and emissions continue in a business-as-usual scenario.[3] The economic costs of these severe weather phenomenon are growing, as are the interconnected, mounting, and costly effects of global warming on ecosystems and countries, notwithstanding Nordhaus's low-ball estimates of damage.

Swiss Re, which has one of the most highly skilled teams of climate risk analysts on the planet, has sounded the alarm perhaps loudest of all. They have estimated the economic costs of climate change, which they must do to avoid huge costs as a reinsurer, and they find the economic costs are very large and could become enormous; sobering findings, indeed.

Ecosystem collapse is not a minor matter

Swiss Re estimates a fifth of countries worldwide are at risk from ecosystem collapse due to a decline in biodiversity. All people in all countries are reliant on their natural ecosystems; we live in and on them. Economies rely on what is termed 'biodiversity and ecosystem services' (BES). These include, in plain English, food provision, water security, and air quality, all of which are vital to the stability of communities and economies. BES degradation would have colossal costs that would be many magnitudes greater than DICE's damage figures.

Swiss Re found that both developing and advanced economies are at risk, and that developing economies that have a heavy dependence on the agricultural sectors (such as India and Pakistan, although the list is long and the populations affected huge) are especially susceptible to BES shocks. The report also found that more than a third of the ecosystems of 39 countries is in a fragile state; that Bahrain, Cyprus, Israel, Kazakhstan, and Malta have the lowest BES ranking; that 55 percent of global GDP depends on high-functioning BES, worth US$41.7 trillion; and that major economies in Southeast Asia, Europe, and the US are exposed to BES decline (Swiss Re, 2020). The report warns:

> Major economies in Southeast Asia, Europe and America that already have diversified economies are nevertheless also exposed to risk from BES decline. This is because important individual economic sectors can be impacted by single BES factors such as water scarcity, which can have a disruptive effect on a country's manufacturing sectors, properties and supply chains.
>
> *Swiss Re, 2020*

If one of the world's leading reinsurers is sounding the alarm, producing estimates vastly larger than DICE, we ought to pay attention. Swiss Re's nightmare BES collapse scenario ought to spur us to action. We can also daily and yearly see the costs mount across the planet, signaling DICE damage assumptions are too low.

Unbearable heatwaves

Heatwaves are increasing in severity and probability, and the heat records being broken today underscore the dangers we face. In 2020, Baghdad, for instance, hit 52 degrees Celsius, a staggering number – too hot to cook a nice cut of salmon properly but just right for a rare steak (*Washington Post*, 2020). Damaging heatwaves are repeatedly being recorded, as in Europe (2003, 2017, 2020); Russia (2010); and East Africa, along with drought (2017); and drought, as in Southern Africa (in 2015). Research suggests billions of people will be subject to heat stress by 2100.

Not only are these increasing heatwaves health disasters, but they are also economically costly, affecting agricultural production, and withering, flooding, and damaging crops. Higher temperatures also cut the nutritional value of crops because as temperatures rise the protein content of crops falls. As local temperatures rise, labour productivity declines, especially in regions like India and Pakistan, where workers are exposed to the sun in the course of their daily work in the fields and on the streets. Increasingly, it will become too hot and too dangerous to work – and not just at midday. Without action, 'it's clear as temperatures rise worldwide, the hottest parts of the world could start to see conditions that are simply too hot for us', according to the UK Met Office (BBC, 2020a). Under the IPCC 'business-as-usual' scenario, urban areas in parts of India and Pakistan could be the first places in the world to experience heatwaves that are so hot they would kill a healthy person sitting in the shade. McKinsey (2020) estimates that some regions of India could face a more than 60 percent annual chance of such heatwaves by 2050 and warns that potentially as much as 19 percent of the global surface might become a barely tolerable hot zone by 2070.

What is the economic cost of this disaster? Do we suppose the poor will stay put and die of heat stroke? No, they will move, triggering a great climate migration magnitudes larger than those being seen today (World Bank, 2018; Lustgarten, 2020). Yet our IAMs and DICE do not include these scenarios in their damage function.

Rising tides and howling typhoons

If populations are not baked, they may be flooded. For instance, in 2020, fully 37 percent of Bangladesh was submerged by floods. The economic and societal cost of such massive floods and the gradual loss of the land on which 160 million people live is immense for Bangladeshis, who face flood-driven ecosystem (Swiss Re, 2020) and possible economic collapse.

IAMs and DICE cannot provide us with answers to these personal, economic, and societal tragedies. Simple global macro models cannot capture the enormity of unfolding regional climate disasters or the long-term corrosive effects of repeated severe weather events. It is far from clear that Bangladeshis will be a great deal better off two or three decades from now. The remorseless rise of the sea, increasingly strong typhoons, disrupted or enlarged monsoons, and other natural disasters suggest that climate-change-driven negative economic effects will be more, not less, damaging.

Flooding and related damage are set to increase. For example, low-lying Ho Chi Minh City could have floods with five to ten times the economic impact in 2050; Bangkok, too, is highly vulnerable to inundation and will face 'endless floods' if action is not taken (*Bangkok Post*, 2019). How should we calculate the potential damage and respond to the dangers these disasters present? Global macro models cannot give us the answer. Politics and planning can.

What of the Caribbean and Central American states, some of which already are failing states, with fraying institutions, refugee crises, and increasing societal violence? It is probable that severe weather events, changes in the local climate, and the effects on these countries' economies (on coffee production, for instance) will undermine their economies. What of stressed Southern African states, with rapidly growing populations, weather and ecosystem crises, and corrupt governments and institutions, unable to manage their economies while simultaneously dealing with droughts. Global macro models cannot give us answers here either. All too often:

> Economic assessment fails to take into account the potential for large concurrent impacts across the world that would cause mass migration, displacement, and conflict with huge loss of life.
>
> *DeFries* et al., 2019: 3

Not only the poor pay for climate denial

The poor will as always suffer the most from the climate crisis, but advanced economies are not unaffected. For example, in the UK, the City of London is seeing the effect of rising seas and tides, with the Thames Barrier having been raised 193 times since its completion in 1982 (Thames Barrier, 2020). The barrier protects 1.3 million people and £275 billion worth of central London property which would otherwise be inundated regularly by high tides. Engineers find that the barrier will be insufficient to protect London beyond 2070. Planners in London are already looking for new solutions for the city's warmer, wetter future.

Destructive floods and hurricanes increasingly hit the US. In 2012, US Hurricane Sandy caused US$62 billion in damage; in 2017, Hurricane Harvey caused US$190 billion in damage. Climate effects and severe flooding risk are being reflected in the US housing market, with costs rising in flood-prone regions. As the *New York Times* (2020) observes: 'More banks are getting buyers in coastal areas to make bigger down payments – often as much as 40 percent of the purchase price, up from the traditional 20 percent – a sign that lenders have awakened to climate dangers and want to put less of their own money at risk'. Those same banks are quickly passing on these at-risk coastal mortgages to US government agencies; the banks do not want to be left holding drowning assets.

This type of coastal flooding driven by extreme weather events is very costly. Of the 246 (as of April 2019) weather disasters since 1980, tropical cyclones have caused the most damage, costing US$927.5 billion total, with an average cost of almost US$22 billion per event. In 2018 and 2019, in the US alone, the cost amounted to US$91 billion and US$45 billion (NOAA, 2019).

The polar vortex that froze much of the United States in February 2021 is another instance of a climate-change-related severe weather event. Most of Texas was snowbound, frozen for days; millions were without power or potable water. Scores died of hypothermia. The insurance costs are expected to be among the highest ever seen in Texas.

A recent additional example in Europe drives home the economic damage from climate change. The 2018 recession in Germany was not only caused by an emission cheating scandal in Volkswagen, it was also severely impacted by extremely low water levels in the Rhine, running through the heart of the export-oriented manufacturing powerhouse to the port of Rotterdam. Here, climate interlinkages are at play. The Rhine is fed by alpine glaciers. As snowfall drops, so does ice melt and river flow. As a German analyst noted: 'A warming climate means that incidents like the low river levels this summer are more likely to occur' (*Business Insider*, 2019). While Germany struggled in 2018 with a trickling Rhine, other locations have increasingly caught fire.

Wildfires scorch the globe

Wildfires rage and economic damage is increasing. The 2020 fires in California, Oregon, and Washington State scorched 26 times the land area as in 2019. Reacting to the devastation, the governor of Oregon stated: 'This is truly the bellwether for climate change on the West Coast. And this is a wake-up call for all of us that we have got to do everything in our power to tackle climate change'. The governor of Washington called the fires 'apocalyptic'; and the governor of California stated: 'the debate over climate change is "over"' (BBC, 2020c). California's 2019 wildfires and the destruction of the town of Paradise cost US$80 billion.

Fires rage elsewhere, as well. In 2019, Australian wildfires, 25 times larger than the previous record fires, cost the country up to US$100 billion and killed 480 million animals and reptiles. In 2018, fires in Canada caused US$10 billion in damage and burned 1.5 million acres. And in 2019, Siberia's 21,000 square miles of tundra went up in flames adding to GHG emissions and signalling what might become a regular occurrence in the melting north (McKinsey, 2020; University of Sydney, 2020).

The Siberian tundra is being scorched. Spring 2020 saw temperatures in Siberia within the Arctic Circle soar to 38 degrees Celsius (100 degrees Fahrenheit), the hottest ever recorded (BBC, 2020b). This was 18 degrees Celsius higher than the average maximum for June. Such record-smashing temperatures underscore that the Arctic is warming at more than twice the rate of the global average, with concomitant effects on forest fires, agriculture, local permafrost melt, and GHG emissions.

As Krugman observes: 'While it will take generations for the full consequences of climate change to play out, there will be many localized, temporary disasters along the way. Apocalypse will become the new normal – and that's happening right in front of our eyes' (Krugman, 2020). As we recalibrate our narratives and our models for this shifting reality, we must recognize the increasing severity of individual weather events and also turn to the even more disturbing matter of fat tails and tipping points.

Fat tails and tipping points: A pessimists' confession

I am a Scot and come from a pessimistic, multigenerational, Calvinist tradition. I am far from the only Scot to be a pessimist; Duncan, for example, recalls in *The Wee Book of Calvin* his grandfather's innumerable negative sayings dished out on a regular basis, including, 'darkness will keek through the smallest hole' and 'if you dinnae expect onything you willnae be disappointed' (Duncan, 2004). I, too, struggle to 'Always look on the bright side of life' (Monty Python, 1979) and to avoid falling back into my Scots pessimism.

Optimism can be quite difficult when discussing fat tails, tipping points, and feedback loops. Yet we must consider the dangers of fat tail risks and nightmarish tipping points.

Sudden leaps and nonlinearity are not an economist's best friend

Fat tail risks are an essential part of our climate change stories and risk calculus, yet many economic models do not include them. This is because of complexity, uncertainty, scale, and timelines that are beyond the framework. The default is therefore to consider smaller, more minor risks that can be extrapolated from gradual shifts seen in past evidence and experience. Economic storytellers, including Nordhaus, suffer from Taleb's (2010) 'turkey waiting for Christmas' problem. We extrapolate forward from the past like forecasting turkeys, missing the fatal fat tail risks of sudden death at the farmer's hand on Christmas eve that is ahead of us.

Fat tail risk is defined as the probability of a risk occurring that is more than three standard deviations from the normal distribution. Figure 2.1 contrasts fat tail risk with normal distribution.

The maximal climate change fat tail risk is of a different magnitude than normal economic or financial fat tail risks because the 'extreme downside is nonnegligible. Deep structural uncertainty about the unknown unknowns of what might go very wrong is coupled with essentially unlimited downside liability on possible planetary damages' (Weitzman, 2011). Wagner and Weitzman (2015) found that the cost–benefit analysis commonly used in economics significantly understates the fat tail risk of climate change. They stress that if governments took these risks into account when making policy decisions, the pressure for action would increase and shift decision-making processes. Wagner and Weitzman (2015) calculated that we face a greater than 10 percent chance of the Earth's eventual warming of 6 degrees Celsius or more – the end of the human adventure on this planet as we know it.

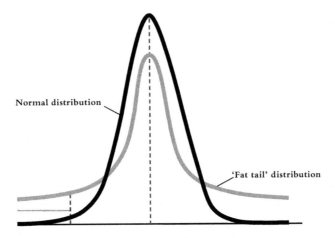

FIGURE 2.1 Normal distribution and fat tails

A civilization-destroying climate change fat tail risk of 10 percent ought to drive us all to immediate action. After all, if I told a friend there was a greater than one-in-ten chance of them being killed walking down their street, they would refuse to walk down it and demand protective action and more policing expenditure. Yet today, many businesses, politicians, and a significant proportion of voters are still willing to take such personal, business, and planetary risks.

Consider the business decisions taken by Pacific Gas and Electric, the bankrupt electric utility in California. In a rapidly warming and drying climate, they knew, or should have known, the extreme fat tail risk posed by the old powerlines and elements of their distribution network that could cause runaway wildfires. The company's cost–benefit analysis should have included those dangers and the possibility of huge liabilities if a fire occurred that torched the landscape and killed people. Yet the firm failed to repair the lines to ensure they could operate safely in a tinderbox environment. Eighty-six people burned to death in the town of Paradise in 2018 because too-rosy, middle-of-the-road, erroneous cost–benefit assumptions were made.

On a planetary scale, we face not one possible fat tail risk but many progressively alarming, even terrifying, interconnected, nonlinear fat tail risks, all tipping points to new equilibria we do not want and cannot deal with. If we fail to act on Wagner and Weitzman's (2015) warning, we will reach the end of our civilization via a series of climate tipping points.

Climate change is slow, then fast

History shows that climate change happens slowly, almost imperceptibly, until it happens very fast, indeed. The archaeological records of the Palaeolithic and Younger Dryas periods are evidence of this (see Box 2.2).

BOX 2.2 SUDDEN SHIFTS IN CLIMATE CHANGE HAVE OCCURRED

The Younger Dryas (14,500 to 11,000 years ago)

The Younger Dryas epoch – named after a flower, the *Dryas octopetala*, which was then common in Europe – lasted from approximately 14,500 years ago to 11,000 years ago. During that period, the Earth's climate shifted rapidly from gradual warming to cooling, to a mini-ice age, to suddenly warming again. Crucially, these transitions between the temperature and climate states occurred very rapidly. Approximately 11,500 years ago, temperatures in Greenland, measured today by analysing ice cores, rose a staggering 10 degrees Celsius in ten years, while this massive shift in other parts of the world occurred within 30 years or less (Alley, 2000).

What this archaeological episode in Earth's history underscores is that change in complex, nonlinear, interconnected climate systems happens slowly, and then change can occur very fast indeed. This is the nature of major shifts in state. Change is not always a gradual, smooth process. It can be a sudden, staggering shift from one state to another equilibrium. When a new equilibrium is reached, there is no going back, at least within timescales relevant to our lives and civilizations.

The Palaeocene warming

The Earth's surface warmed by approximately 5 to 8 degrees Celsius during the PETM 55 million years ago. The PETM was one of the most rapid and dramatic instances of climate change in Earth's history (Wright and Schaller, 2013).

The warming resulted in most ice sheets melting; this coupled with thermal expansion of ocean water and other factors meant sea levels were approximately 70 to 140 metres higher than they are today (Haq *et al.*, 1987) for a duration of over 15,000 years. This is a slower rate of sea level rise than is occurring at present.

The sea levels rose at a rate of between 1 and 2 metres a century – which would be fast enough to force the repeated relocation of major ports if it occurred today and result in trillions of dollars in real estate losses (a majority of the world's population live in sea level coastal conurbations).

Sea temperatures during this time were much warmer than they are at present, with temperatures off the coast of Antarctica estimated at 20 degrees Celsius. The oceans of the tropics were bath-like. Temperatures off the coast of West Africa were 36 degrees Celsius (97 degrees Farenheit). Such temperatures and ocean chemistry would have been fatal for many sea creatures and fish because of the destruction of their shell structures (Earth in the Future, 2021).

> The PETM period might seem distant and the rate of change slow, but it was not. In the PETM period, the Earth warmed greatly twice (called hyperthermals). Those instances of warming took place in as little as 13 years (Wright and Schaller, 2013). The mechanisms are unclear, but the message is not. Once the shift occurred, there was no going back. Temperatures remained elevated for 170,000 years (BBC, 2015).
>
> Both these cases from prehistory should remind us all that climate change appears slow, until it is not.

Much about our planet appears relatively stable and the climate changes almost imperceptible. Yet, between visible disasters, small changes mount, add up, and increase the stock and probability of increasingly severe sudden outcomes. Even 'little things make a difference' (Gladwell, 2000: 1), especially when multiplied billions of times. Eventually, and probably sooner than we estimate, our planetary systems may be switched 'into a qualitatively different state by small perturbations … the tipping point is the corresponding critical point – at which the future state of the system is qualitatively altered' (Lenton *et al.*, 2008).

Scientists recognize that tipping points are composed of a series of interconnected feedback loops and spirals we only partially understand. But they observe that the feedback loops draw interrelated climate and physical tipping points closer and faster and increase local and global temperatures. Some locations have already seen regional tipping points being reached.

Some locations already have experienced tipping points

Australia's fires (and California's in 2020) show us fat tail events; tipping points to a new equilibrium may not only occur at the far end of the distribution. Australian policymakers gamed climate change risk scenarios in the spring of 2019. They identified the key fat tail scenario of high-temperature drivers, drought, high winds, and the result of a series of major bushfires all down the South East Coast. That fat tail risk played out just months later. Australia has quite possibly already crossed a tipping point to a permanent new wildfire equilibrium. No one expects the country to suddenly get wetter, stop experiencing severe droughts, see bush fires decline, and return to the more liveable climate of the 1940s and 1950s. Similarly, California and parts of the US West may also be facing their own tipping point in terms of the scale of yearly conflagrations and destruction.

Figure 2.2 shows how fat tails and tipping points might be interconnected. Australia's 2019 fat tail event and Tipping Point A moves the country to New Equilibrium A, a situation where severe wildfires are the norm. A tipping point can be reached when all appears otherwise gradual. We cannot assume linearity or gradualism in climate change. Australia is not alone.

The tipping point for small island nations in the Pacific is here now. They face literal submersion of their homes. Kiribati, an island nation in the mid-Pacific that

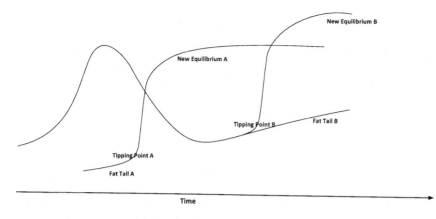

FIGURE 2.2 Tipping points and new equilibria

gained its independence from the UK in 1979, is an example. The country is only 2 metres above sea level at its highest point, and Kiribati is drowning. One in seven of its population are already climate refugees. You cannot say to Pacific Islanders: 'Do not worry, we haven't reached a tipping point and crisis yet … hang on'. They cannot; the tipping point for them has arrived (BBC, 2019). For other regions, a tipping point is more likely to occur later, as Figure 2.2 suggests, represented by Fat Tail B, Tipping Point B, and New Equilibrium B.

When a tipping point – a juncture of sudden change – occurs, there 'are physical processes acting as positive nonlinear climate and biosphere feedbacks that, after passing a threshold, could irreversibly shift the planetary system to a new warmer state' (Steffen *et al.*, 2018). I noted at the beginning of this chapter that these tail risks do not fit easily into our narratives, let alone our economic stories and models, as they are truly 'unprecedented in human history' (DeFries *et al.*, 2019: 12). Yet we nonetheless need to understand them and draw policy and personal conclusions.

Appendix 2.1 Boxes A2.1–A2.7 address tipping points in ascending order of planetary climate danger and temperature rise. In each title, I have placed the current scientific tipping point estimate for the temperature increase required to irreversibly trigger the process in question. This is not an exhaustive list; there are others that are known but most are only partially understood, and others that are still to be discovered. Scientists are constantly finding new interlinkages and complexities they did not know about or only partially understood.

The timescales for these tipping points and shifts in climate state or equilibrium can be very long. Many are often way outside forecasting ranges. Others are visibly getting nearer, hitting new records, year after year.

Tipping points do not make for enjoyable reading. Some are already upon us. Others loom large if we fail to act 'right here, right now' (Thunberg, 2019). Tipping points must be part of our climate crisis economic calculus and policy matrix.

If tipping points upset us, so should climate sensitivity. Climate sensitivity is the extent to which the planet's systems respond to a given temperature increase. It underpins all the IPCC analyses and models. If the planet is less sensitive, this is good news, for we will have more room to act. If the planet is more sensitive than we have assumed, the news is dreadful because disaster is closer to us. Box 2.3 discusses growing evidence we may have climate sensitivity wrong.

BOX 2.3 ESTIMATES OF CLIMATE SENSITIVITY MAY BE TOO CONSERVATIVE

Our planetary diagnosis – our measure of the severity of the interconnected climate illness – may be too optimistic. Researchers suggest we may have got one essential model driver – climate sensitivity – wrong. If this is the case, the need for urgency and action is even more imminent. Climate sensitivity refers to the assumption made in IPCC models regarding the effect of a doubling of GHG in the world's atmosphere over preindustrial levels. IAMs and DICE rely on consensus assumptions regarding climate sensitivity. But these are not static. Work done in 2020 is signalling we may be underestimating this crucial factor.

Until now, the models have generally estimated that a doubling of GHG over preindustrial times could result in up to 3 degrees Celsius of warming, giving policymakers and actors still some room for manoeuvre towards a glidepath to net zero and to aim for a 1.5-degree Celsius rise as per the Paris Agreement, although the glidepath is steepening and narrowing. But what if climate sensitivity is underestimated, not just by the IPCC modellers but by many other models? In 2020, modelling from institutions that are part of the sixth assessment IPCC report process, due to be released at the COP26 meeting in Glasgow in November 2021, indicates that 25 percent of models show a sharp shift in sensitivity upward from 3 degrees to 5 degrees Celsius (*The Guardian*, 2020b).

A quarter of the best models we have now estimate the possible upward effect of GHG doubling at as much as 5 degrees Celsius. These models are not built by cowboys. They include the UK Met Office and the European community's Earth-System Model. Perhaps the models are wrong. But if there is even a modest possibility such estimates are correct, we all confront the real and present danger of triggering multiple tipping points sooner and a concomitant leap to a new, frightening and dangerous hotter climate equilibrium, even if we hit current targets.

If we have climate sensitivity wrong, the following tipping points could be reached within the lifespans of our children: loss of Arctic summer ice, loss of Arctic glaciers, collapse of the Greenland ice sheet, collapse of the West Antarctic ice sheet, dieback of the Amazon, and dieback of the boreal forests.

Given the scale on which these interconnected planetary processes operate, there is a need to acknowledge a great deal of uncertainty. We still do not know the precise effects of one system on another. The complexities are hard to understand and confounding. In many subject areas, we have only very partial knowledge. These tipping point scenarios are still just that – scenarios – but they are constantly being tested by the world's scientists, probed, and added to with more data and adjusted models. Nonetheless, over time, the direction and interlinkages are becoming better understood and backed by more empirical data. The directional trend is consistently towards greater warming and greater dangers of reaching a tipping point or points.

Climate crisis economics must include local tipping points and global fat tails

Climate crisis economics narratives and scenarios should include low-probability, high-risk events in their models. Policymakers should – as banks and insurers and reinsurers now do – stress-test plans and strategies against low-probability, extremely high-risk, high-cost outcomes.

Too often our economic stories and models do not include fat tails, because it is easier to exclude complex and low-probability events. This is wrongheaded. It is these gray rhinos (Wucker, 2016) – known but ignored climate change risks and events – that might kill you if you live in Bangladesh, New Orleans, Southern California, or the Caribbean.

Our models should in the future include national and global fat tail risks. Many sector-specific models in insurance already include extreme weather events, cyclones, hurricanes, and droughts. These risks need to be planned for and assessed. As Ward states:

> Economists and finance ministers must stop relying on models that are simply not fit for purpose … the potential impacts of climate change caused by fossil fuel use are grossly underestimated by the current generation of economic models, which cannot quantify the cost of, and therefore omit tipping points in the climate system.
>
> *Ward, 2018*

Just as the IPCC models are under continued refinement, adjustment, maintenance, and replacement, so our climate change economic models need to reflect shifts in scientific understanding and adjust accordingly. If our models fail to adequately capture reality, we need to discard them and design alternative ones that are better suited to the task.

When the DICE IAM concludes the 'optimum' outcome is a temperature rise of 3 degrees Celsius by 2100 or 4 degrees Celsius by 2150 (Nordhaus, 2018), far, far above that agreed as the maximum by almost all governments on the planet, this is distortive and damaging, not helpful (Ward, 2018).

If a model fails to help us illuminate and achieve the agreed climate change goals required to assure human survival and instead the model's assumptions act as a brake on action, we need to look elsewhere to support effective policymaking and planetary survival.

If your doctor misdiagnosed you repeatedly, what would you do? Suppose you had a slow-burning, gradually rising fever and you went to one doctor and they told you: 'Don't worry. Right now, it's not worth doing anything. Just carry on as normal'. Perhaps they might also add, frustratingly, 'sorry I can't fully diagnose you anyway; your symptoms are too complex and wide ranging'. You leave but return a few days later. The fever continues and is now slightly higher. Yet again, you get the same response, with the additional surprising observation that: 'No, don't spend money on prevention or treatment, go out and enjoy yourself, live for today not tomorrow'. This happens again on your next visit, and still the next, albeit with some – unhelpful – expressions of general concern. Would you stay with the same doctor or look for someone else who could diagnose you and treat the ailment, even if it meant spending money today to make you better tomorrow and the next day? What you would do is self-evident: You would change doctors and get a proper diagnosis and a plan of care and recovery, even if costly in the short term. If you make a mistake and are dead tomorrow, having saved today and failed to treat what ailed you with a treatment that might have cured you, is extremely foolish indeed.

Time to act, test models, and use all available policy levers

To repurpose a Covid-19 phrase, societies must work to flatten the climate change curve (as per Figure 2.3).

DICE and its narrative face justified criticism because the assumptions used and the key conclusion produced a misguided conclusion – not to spend now to

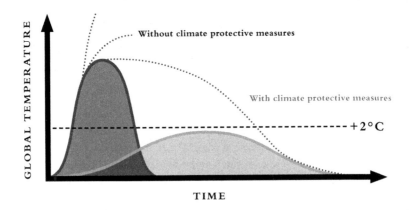

FIGURE 2.3 Flattening the climate curve

Source: https://flattentheclimatecurve.org.

mitigate disaster tomorrow. DICE's discount rate is still too high today. This acts as a brake on policy action. In addition, the DICE carbon price remains too low. In 2018, the optimal carbon price recommended by Nordhaus was US$36. This price is not high enough to help achieve the Paris Agreement goals and thus fails to help protect the planet. As we have seen, other assumptions – on the cost of mitigation over time, economic growth rates over time, and estimates of the damage of climate change – are also open to doubt.

What one might call the DICE denial story has dominated the economic policy debate and affected the policy dialogue negatively, lulling the audience into a false sense of security, and then slow-peddling the building severity of what we face and must respond to globally. Hickel (2018) places the blame squarely on DICE: 'The failure of the world's governments to pursue aggressive climate action over the past few decades is in large part due to arguments that Nordhaus has advanced'.

DICE has many critics, but their warnings have not received adequate attention. Stern (2006), Stiglitz (2019), Weitzman (2011), and others have argued strongly for different assumptions on discount rates, carbon pricing, fat tails, and tipping points which lead to different outputs and increase the urgency for policy action. Unfortunately, the voices of alarm have been outmatched by the ideological use of a neoclassical model that has played into a denialist narrative of relative inaction that resonated among politicians who resist acting robustly across sectors and tax policies. When needed, they could turn to DICE as justification for delay and a reason not to invest today for their own children's tomorrow.

In the end, as Parramore (2019) states, a model is only useful when it assists in achieving our goals:

> The point is not to endlessly debate, for example, the optimal social cost of carbon, acting like medieval scholars arguing how many angels can sit on the head of a pin. Efficiency is not the issue. What matters is how effective the climate change policy is going to be.

Climate crisis economics must take this admonition to heart. The operative question is whether policy A, lever B, or tax C supports and speeds us towards the net-zero goal before climate catastrophe. Models, policies, and approaches need to be measured against this first and foremost. Other factors can be considered, political calculus adjusted, but all policies and processes must be aligned with and embedded with the global, regional, and local climate change goals, their targets, and deliverables.

Economic efficiency is important but cannot be the sole determining factor. Survival of humanity, the Amazon, Siberian forests – indeed, of all life on precious Planet Earth – is not just a matter of economic efficiency. The stakes are so much greater than that. To succeed in the management of the climate crisis, governments and communities need to use every possible lever, mechanism, and incentive to achieve net zero, including markets, trading, taxes, incentives, penalties, tariffs, mandates, and prohibitions.

Can we afford it? Where will the money come from? I hear those questions often, but they are the wrong questions. When you are in a war for planetary survival, you do not ask: Can we afford it? You take up your arms. You ask: What do I need to do to help win the war against carbon and GHG? (*Forbes*, 2020; Stiglitz, 2019). It is a matter of survival. We have no alternative but to take up economic, regulatory, and societal arms against the rising tides, fires, and dangerous distant and not-so-distant tipping points.

At present, most markets do not fully internalize the cost of GHG emissions and carbon. Markets will not do this until governments change the policy narrative permanently, through taxes and regulation. At COP26, leading countries must take major steps to set a minimum and gradually raise the price for carbon and regulate the global and regional carbon markets, such as the EU Emissions Trading System (ETS) market in Europe, the new Chinese markets, and the fractured and patchy US cap-and-trade market.

Leaders at COP26 and in the years ahead must harness markets to achieve net zero. They must speed the creation of markets for offsets, which must be grown rapidly. As markets are yoked to climate change goals, state power must help ensure we achieve net zero.

In addressing climate change, we are making not only economic but also political and ethical decisions. This is necessary and appropriate. Solving for climate change is so much greater a challenge than determining the utility of a single action by an individual or group. We need to bear economics in mind but extend our framework to moral and ethical planetary considerations, as well.

Perhaps we can look for advice from the 'father' of economics, Adam Smith, to help us think more broadly, inclusively, and outside of models, and include within our stories and narratives matters of ethics and morality. Smith focused not only on the actions of the 'invisible hand' and individuals in the Industrial Revolution but was also aware of and concerned about the moral underpinnings of decision making and actions. We would do well, as we consider how to construct climate crisis economics, and our models, to draw upon these economic drivers but also on the ethical and moral considerations Smith raised.

A return to Adam Smith as moral philosopher

Adam Smith was no market ideologue. Rather, he was a balanced thinker who was aware of the limits of the discipline he was instrumental in creating. Smith was, first and foremost, a moral philosopher. He sought to understand and reconcile the different drivers of human action and decisions. *The Wealth of Nations* (Smith, 1776) explained how the selfish individual in business can support society without ever intending to do so:

> Every individual necessarily labours to render the annual revenue of society as great as he can. He ... generally ... neither intends to promote the public

interest, nor know how much he is promoting it .... [H]e intends only his own gain, and he is in this, as in many other cases, led by an invisible hand to promote an end which was not part of his intention.

Smith, 1776: Part IV, 2.9

Smith brightly illuminated the operation of markets at the time of the British Industrial Revolution. He described how mechanization could increase productivity and output for the owner and the worker in pin factories and elsewhere. His insights were sharp, and his conclusions have driven and informed economic thinking ever since. Many economists, however, have latched onto *The Wealth of Nations* and misrepresented its contents. As Fosler (2013) notes:

> The more libertarian factions in the economics profession have hijacked Adam Smith as the progenitor of economic liberty and the efficiency of the modern market economy. As such, his name is often invoked to argue against government intervention in society and markets and in favor of freedom of market-determined outcomes.

Such a selective reading does not properly reflect Smith's wide-ranging thinking on morality and economics. Smith did not miss the negative dynamics that could arise in markets and industries. He was far from a naïve cheerleader for markets and their operation. He saw that markets do not always operate to the benefit of society at large. He knew industrialists and business owners would distort markets if given the opportunity, observing:

> People of the same trade seldom meet together, even for merriment and diversion, but the conversation ends in a conspiracy against the public, or in some contrivance to raise prices.

Smith, in Sagar, 2018

Smith's clarity on the potential negative effects of selfish acts by individuals makes his explanation of the economy more nuanced and complete than is allowed by today's market fundamentalists, who hold tight to what they agree with on markets (in Smith, in Friedman (1970), in Hayek (1943). They overlook Smith's most influential text, *The Theory of Moral Sentiments* (1759), which too few read or remember. In *The Theory of Moral Sentiments* we see that Smith was not a market fundamentalist. Rather, he drew on many strands and ideas to understand human nature. Today, unfortunately, 'the misrepresentation of Smith's writing has had a persuasive and regrettable influence' (Graafland and Wells, 2019).

In his lifetime, Smith was often highly and persistently critical of the 'commercial' men of industry – the greedy, the self-interested, the desirous of riches. Smith did not laud this grubby conduct. Instead, Smith worried about those who became besotted by business, warning that they may become 'stupid and ignorant', stating:

> The torpor of his mind renders him, not only incapable of relishing or bearing a part in any rational conversation, but of conceiving any generous, noble, or tender sentiment.
>
> *Smith,* The Theory of Moral Sentiments, *in Rasmussen, 2017: 171*

Friedman, Hayek, and Rand (1957) laud the mythical free market. Smith did not do that. His work is full of passages lamenting the potential moral, social, and political ills of what he called commercial society, by which he meant those driven only by personal self-interest. Not only did Smith warn of the intellectually corrosive effects of markets, he also warned of the impact of inequality and the division of labour. Smith wrote:

> the labouring poor, that is, the great body of the people, must necessarily fall, unless government takes some pains to help prevent it.
>
> *Smith, 1776: Book V, Chapter 1, Part 3*

This is Smith as a supporter of government regulation and intervention. Smith knew and worried that:

> Wherever there is great property there is great inequality. For one very rich man there must be at least five hundred poor, and the affluence of the few supposes the indigence of the many.
>
> *Smith, 1776: Book V, Chapter 1, Part 2*

Smith was worried about the ill effects of extreme inequality.

Smith thus thought broadly, concerned by the failures of markets and commercial society. He sought to support virtue and was sympathetic to the plight of his fellow human beings. Why is this important? Because as we seek to construct pathways to net zero, we all need to see markets as they operate. We need to recognize when they do not work and address the failures. We must take a moral stance and think about markets serving society, as part of a moral and ethical system, not separate from it. To address climate change, we must reach beyond the purely economic, to be willing to recognize the need for moral and ethical judgements, to consider what is right and just not only what has the greatest efficiency or the greatest short-term utility.

We need a *homo economicus sympatico*

I suggest we reach beyond a stunted *homo economicus* to an alternative hybrid species, which I call *homo economicus sympatico*: An actor that balances the economic with morality, ethics, virtue, and sympathy for others, much as Adam Smith did (see Box 2.4).

BOX 2.4 WHO IS *HOMO ECONOMICUS SYMPATICO?*

To respond to this epochal climate change challenge, we need a new avatar to represent the struggle we are engaged in, a personification that is both economic and ethical – a *homo economicus sympatico* (HES)[a] – an economic actor but not a solely, coldly utilitarian one.

HES acts on economic drivers such as pricing, incentives, and penalties, hence the need to design climate change policies to trigger that part of HES's nature. But HES also has an evolutionary tilt towards collaborative, communitarian solutions as well as individualist responses. HES has a clear sense of what is fair and what is not, for themselves, and for others, and will act on that evolutionary driver.

HES is an ethical actor who considers not only their own short-term daily wants and desires but also the needs of HES's family, community, and future generations. HES is forward looking and can act today to prepare for tomorrow.

HES is an economic and empathic actor and considers the impact of their actions on others. HES has sympathy for others and is concerned how others view their actions and decisions.

HES as a moral actor considers the price of a good or action but also the ethical cost.

HES ponders the value of items without explicit pricing.

HES can place non-monetary moral and ethical value on other species and inanimate objects – the air, the soil, the sea, the rivers – that HES relies on for physical, psychological, and emotional survival.

In summary, HES is an economic and moral actor and can make decisions based not only on economic grounds but also considering other ethical aspects.

HES is much closer to an actual description of humanity today than their intellectually and ethically challenged predecessor.

Note: [a] I am aware that *homo economicus sympatico* is a mixed bag of linguistic nomenclature. Nonetheless, I think it aids the reader in envisioning how such an individual behaves. This avatar is not solely driven by selfish economic drivers today, disregarding tomorrow and their communities' and families' survival. Indeed, all too often our economic analyses can be stunted by adherence to wrongheaded assumptions about motivations, behaviour, and actions. Empirical evidence in behavioural economics and neuroscience increasingly shows we have more complex, communitarian, and altruistic reasons for acting. This is in fact good news because it opens up the possibility for new narratives of action, cooperation, and survival that the old *homo economicus* would not permit.

I will return to this hybrid species later in the book.

For now, we need to turn to the crucial matter of pricing carbon in our markets and internalizing the cost of pollution, and in doing so hasten the achievement of net zero.

Chapter 3 addresses the key failure in economics and the economy vis-à-vis climate change, namely the persistent failure to internalize and price carbon in the economy and in our markets. It looks at how we can price carbon effectively, what mechanisms work, what lessons can be lifted from past and current practices, and how we can proceed to price carbon, harness the markets, and help ensure our collective net-zero goal is brought forward and climate tipping points are avoided.

Appendix 2.1

BOX A2.1 LOSS OF ARCTIC SUMMER SEA ICE – ALREADY UNDERWAY – 1 TO 2 DEGREES CELSIUS

One of the most visible tipping points, which is already underway, is the loss of Arctic summer sea ice. IPCC models show that since the 1970s, the rate of ice loss has accelerated and exceeded model forecasts every year. Measurements in 2019 suggest that a total loss of summer sea ice in the Arctic is now likely before 2050, even if GHG emissions are cut rapidly (*Geophysical Research Letters*, 2020). As one of the study's authors stated:

> Alarmingly the models repeatedly show the potential for ice-free summers in the Arctic Ocean before 2050, almost irrespective of the measures taken to mitigate the effects of climate change. ... The signal is there in all possible futures. This was unexpected and is extremely worrying.
>
> *The Guardian*, 2020a

Arctic sea ice, which reflects sunlight and heat back into space, has been thinning and shrinking for years. Data from satellite records show that since 1979, summer Arctic ice has lost 40 percent of its area and up to 70 percent of its volume, making it one of the clearest signs of human-caused climate change. In 2019, the ice sheet shrank to its second-lowest coverage on record.[a] Failure to cut GHG as required by the COP process will guarantee Arctic summer ice will vanish permanently. If GHG emissions remain high, there is a risk the Arctic could be ice-free even in the dark, cold winter months, a frightening possibility. As Wadhams states:

Sep 15 2020

FIGURE 2.4 2019 Arctic summer sea ice at second lowest level ever

Source: NASA. See www.climate.gov/news-features/understanding-climate/climate-change-minimum-arctic-sea-ice-extent.

> The great white cap that once covered the top of the world is now turning blue – a change that represents humanity's most dramatic step in reshaping the face of our planet.
>
> Wadhams, 2016

The melting ice cap shows climate change is underway and accelerating, as Figure 2.4 shows.

The melting also demonstrates that the process of change appears gradual until suddenly a disjunctive break occurs, and a new state is reached. For example, in late July 2020, the last Arctic ice shelf in Canada disintegrated. The shelf was 80 square kilometres – bigger than the island of Manhattan. Over a period of only one week, the entire shelf broke up. This icy tragedy illuminates how tipping points play out – slowly, and then a sudden disjunctive break occurs.

The practical certainty of a loss of summer Arctic sea ice is the clearest signal that warming is underway and causing feedback loops. The ice cap is in a death spiral feedback loop. This is a disaster both for polar bears and for the indigenous peoples living around the Arctic Circle. The ice loss is also contributing to increased extreme weather events (so called polar vortexes), along with increased heatwaves in the summer.

Note: [a] This 2019 Arctic ice research is based on 40 of the latest computer models. It is viewed as the most accurate current assessment of the fate of the Arctic ice.

BOX A2.2 LOSS OF ALPINE GLACIERS – ALREADY UNDERWAY – 1 TO 2 DEGREES CELSIUS

The Earth's alpine glaciers are melting fast. The year 2018 was the thirtieth consecutive year of their shrinkage. The decline is unprecedented (Zemp *et al.*, 2015). The glaciers, one of the most sensitive indicators of climate change, will soon be gone (Huss, 2017). As Pelto (2019) notes: 'Just as people need to consume as many calories as they expend or they will lose mass, glaciers need to accumulate as much snow and ice as is lost to melt and calving icebergs in order to survive'. Thirty consecutive years of alpine glacier mass loss heralds the end of alpine glaciers.

As the world warms, less snow falls, less ice forms, and existing ice melts. The thinning, retreat, and shrinkage of the world's 2,000 alpine glaciers is consistent across all continents and regions and is directly caused by human GHG emissions in the Anthropocene (Marzeion *et al.*, 2014).

The loss of glaciers has serious consequences for the billions of people that rely on the gradually shrinking flow of water from rivers that emerge from these mountain ranges. The start of this tipping point is underway, and we appear unable to stop it. As we have seen, the Rhine can almost run dry. The Mekong can shrink. The economic and societal effects are real, damaging, and huge. As the glaciers shrink, the planetary and economic costs will only rise. The Estonterm Glacier in Washington State is melting fast, as shown in Figure 2.5.

FIGURE 2.5 Alpine glaciers are disappearing

Source: Wikimedia Creative Commons Easton Glacier on Mount Baker in retreat, public domain. See https://fr.m.wikipedia.org/wiki/Fichier:Eastonterm.jpg.

BOX A2.3 COLLAPSE OF THE GREENLAND ICE SHEET >2 TO 3 DEGREES CELSIUS

The Greenland ice sheet is melting. The summer of 2019 was exceptionally warm, with record temperatures. Ice cap melt was recorded across 90 percent of the island (NSIDC, 2019). The clear and sunny weather in Greenland in the summer of 2019 saw the second-highest amount of runoff from melting ice ever (2012 was worse) (see Figure 2.6); the so-called 'surface mass balance'[a] fell 320 billion tons, the steepest fall since recordkeeping began in 1948 (Fecht, 2020).

Scenarios – ones that do not assume a complete collapse of the ice sheet – suggest that the Greenland ice sheet melt will be at between 2.7 and 12.9 gigatons per year, with ice slabs, which speed melting dynamics, affecting an area ranging from 334,000 to 610,000 square kilometres (McFerrin *et al.*, 2019). So far, this melt has added less than 1 millimetre in global sea level rise, but if the melt proceeds as forecast, even if we avoid triggering its total collapse and limit it to 'only' a considerable retreat, the impact will be much larger this century, adding up to between 50 centimetres and 1 metre to global sea level rise.

FIGURE 2.6 Greenland ice sheet melt

Source: Wikimedia Creative Commons, CC By 2.0. See https://commons.wikimedia.org/wiki/File:Greenland_Ice_Sheet.jpg.

Note: [a] Surface mass balance includes gains in the ice sheet's mass, such as through snowfall or losses from surface meltwater runoff.

BOX A2.4 COLLAPSE OF THE WEST ANTARCTIC ICE SHEET >4 TO 5 DEGREES CELSIUS

The Antarctic ice sheet represents the largest potential source of future sea level rise. That is, if all its ice melted, the sea level would rise by about 60 metres. Scientists worry particularly about the instability of the West Antarctic ice sheet. According to theoretical (Schoof, 2007) and recent modelling results, this region could be prone to rapid ice breakup. Satellite observations show this is already being seen in the Amundsen Sea, where some of the fastest-flowing glaciers on Earth – the Pine Island (Favier *et al.*, 2014) and the Thwaites Glacier (Jougin, Smith, and Medley, 2014) – are found.

Models suggest that as the West Antarctic ice sheet melts faster and the grounding line[a] retreats, seawater and meltwater can lubricate the flow of ice into the sea, into floating shelves of ice, which are then unstable[b] and can break off, into huge icebergs. For example, in 2019 a 315-billion-ton, 1,600-square kilometre iceberg, the size of Delaware or the Isle of Skye, broke off and floated free (BBC, 2019) (see Figure 2.7).

According to satellite observations, the Antarctic ice sheet lost 1,350 gigatons of ice between 1992 and 2011, equivalent to an increase in sea level of 3.75 millimetres spread evenly across all the world's oceans. This may sound

FIGURE 2.7 Antarctic shelf caves iceberg the size of London, 2021

Source: NASA. See https://visibleearth.nasa.gov/images/148009/breakup-at-brunt.

small, but this rise is over a surface of about 360 million square kilometres. At present, the Antarctic ice sheet contributes about 10 percent of total GHG warming sea level rise, which is less than that coming from the expansion of warmer water (40 percent), glaciers (25 percent), and the Greenland ice sheet (17 percent).

Scientists do not know if or when the melting process will accelerate. The timescale is very long, extending beyond 2100, assuming no catastrophic tipping point is reached before then. But the dangers should this GHG-fuelled melt rate speed up are such that it should underscore the need for urgent action now.

Notes: [a] The grounding line is the line where the ice ceases to be in contact with and faces friction from the ground below. Behind the grounding line the ice moves more slowly. Beyond the grounding line, the ice flows faster, as it is lubricated by the water beneath the base of the ice sheet.
[b] This phenomenon is known as the 'marine ice-sheet instability mechanism', or MISI.

BOX A2.5 DIEBACK OF THE AMAZON RAINFOREST >4 DEGREES CELSIUS

Scientists warn of an abrupt large-scale shift of the Amazon rainforest caused by climate change, known as Amazonian forest dieback. If a sudden withering of the forest occurs, triggered by climate change feedback loops, the tragedy would be staggering. Research conducted since the last IPCC report suggests that a failure to mitigate the possibility of Amazonian forest dieback could result in losses of US$957 billion at the low end, rising to as high as US$3.6 trillion over a 30-year period. Mitigation, by comparison, is estimated at 'only' US$64 billion, while adaption measures taken, rather than acting to halt deforestation, would cost US$122 billion (Lapola *et al.*, 2018).

Despite this, the forest already faces immediate threats. The Amazon rainforest is daily under assault and is being damaged, affecting its ability to breathe. Research suggests the forest's ability to act as a carbon sink has declined by 30 percent over recent decades (Brienen *et al.* 2015). Other work indicates droughts are now more common in the Amazon region. Up to 70 percent of the Amazon rainforest may be subject to dieback and its replacement by savannah by the late twenty-first century if nothing is done (Cook and Vizy, 2007).

The forest fires of 2019 and 2020, the highest number in more than a decade, remind us the forest remains at risk from government failures; the short-term rush to exploit the trees, flora, fauna, and land; and an unwilling-ness to recognize the global public good that is the Amazon rainforest. It may be lost, perhaps not tomorrow but within our grandchildren's lifespans, if we do not act.

BOX A2.6 DIEBACK OF THE BOREAL FOREST >4 DEGREES CELSIUS

The boreal forests, which circle the Arctic, are one of the largest stores of carbon on Earth. A warming planet is bad news for the boreal forests of North America, Scandinavia, and Siberia. Warming increases droughts and fires and adversely affects most species in boreal forests (Olsson, 2009), increasing stress and tree deaths due to disease. Climate models already assume a 2-degrees-Celsius warming in these regions due to GHG. In some areas, very high localized temperature increases of 0.5 degrees Celsius per decade are already being seen.

Models estimate climate zones shifting northwards at a rate of 5 kilometres per year as a result of this warming. This is ten times faster than trees can grow in the northern reaches that may become tolerable as the warming southern reaches become inhospitable for them (Olsson, 2009). We do not live in Tolkien's Middle-earth. The boreal forest trees are not Ents. They cannot uproot themselves and move. As the boreal forests warm and the trees die, huge summer fires are set ablaze by lightning and firehawks[a] (Ackerman, 2020) and the fires rage over increasing parts of the northern landscape. In 2019, hundreds of wildfires burned across Siberia, consuming over 5.9 million acres of woodland (NASA, 2019). The number of fires is expected to more than quadruple as the climate warms (Harvey, 2020).

Once again, the tipping point here will be nonlinear. 'The most likely scen-ario for the boreal forest is a nonlinear response to warming, resulting in the creation of hitherto unseen grassland ecosystems and the extinction of species with limited capacity to adapt' (Olsson, 2009). If we breach the 2-degree limit, extensive decline of these forests is to be expected. If we fail to arrest warming but instead allow temperatures to increase by 3 to 5 degrees, large dieback will occur and a tipping point of terrible consequence might be reached, with feed-back loops spurring faster runaway warming and a release of the carbon stock of these vast forests into the atmosphere. The IPCC warns this scenario is possible.

Note: [a] Firehawks are raptors native to Australia that use fire to smoke out prey. See https://wildlife.org/australian-firehawks-use-fire-to-catch-prey for a fascin-ating description.

BOX A2.7 ACCELERATED PERMAFROST MELT >9 DEGREES CELSIUS (OF LOCALIZED WARMING)

For tens of thousands of years, grasses, other plants, and dead animals have been frozen in the Arctic ground across more than 10.5 million square kilometres. The permafrost is a colossal carbon storeroom that is waiting to be unleashed when the ground thaws. Permafrost melt is one of the biggest tipping points in climate change. Unfortunately, this vast frozen tundra could be reduced to as little as 1 million square kilometres by 2100 if climate change is not arrested.

Data show local warming is more severe at northern latitudes. For example, temperatures in Yakutia,[a] an area one-third the size of the United States, are 3 degrees higher than in preindustrial times. This rapid warming is spurring localized permafrost melt, which is destroying arable land and killing the animal species that local populations rely upon. The Siberian tundra and permafrost are especially vulnerable to self-sustaining collapse, which could release 2.0 to 2.8 gigatons of carbon per year (Troianovski and Mooney, 2019). Research suggests that if nations fail to make a rapid transition to net-zero emissions, the eventual economic damage from widespread permafrost melt could approach US$70 trillion (Yumashev et al., 2019).

The impact of a triggering of permafrost melt in the north would not be confined to the north. Yet, in 2020, the IPCC models did not include the effects of permafrost climate feedback (PCF). Scientists view this as one of the largest and most damaging feedback loops. The timing of the melt is unclear and long term, extending beyond 2100. But localized accelerated permafrost melt has already begun in some regions, including parts of Siberia and Alaska. There is no room for complacency. What governments decide and act upon as a result of COP26[b] may determine whether we ultimately hit a PCF trigger and climate and economic catastrophe. As one observer remarked: 'It is very much a question of "when" [PCF happens] unless we get a grip on climate change very quickly' (*Inside Climate News*, 2019).

Localized permafrost melt is already taking place. It is visible in the profusion of small circular ponds and lakes that dot the Siberian tundra, which are created as the permafrost thaws, as shown in Figure 2.8.

Notes: [a] Yakutia is a federal Russian republic. It was historically part of Russian Siberia but is now officially known as the Republic of Sakha.
[b] COP26: The Conference of the Parties meeting that will take place in November 2021 in Glasgow, Scotland. It may prove to be a pivotal moment in the policy response to climate change. With net-zero commitments from

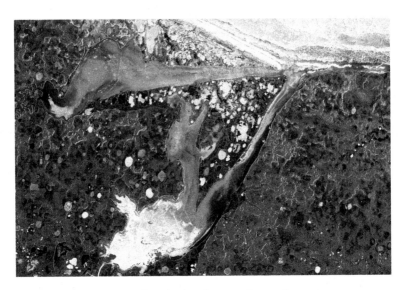

FIGURE 2.8 A landscape pockmarked with permafrost melt

Source: Jesse Allen and Robert Simmon, NASA Earth Observatory, Public Domain. See https://commons.wikimedia.org/w/index.php?curid=16097645.

the US, China, Japan, and others, coupled to market shifts already underway, COP26 is an opportunity for governments to signal a permanent shift in climate policy consensus and underscore the urgency of action.

Notes

1 'Animal spirits' was a term coined by Keynes to describe how people arrive at financial decisions, including buying and selling securities, in times of economic stress or uncertainty.

2 *The Stern Review on the Economics of Climate Change* is a 700-page report released for the Government of the United Kingdom on October 30, 2006, by economist Nicholas Stern, chair of the Grantham Research Institute on Climate Change and the Environment at the London School of Economics (LSE), and chair of the Centre for Climate Change Economics and Policy at Leeds University and LSE. The report discusses the effect of global warming on the world economy. Although not the first economic report on climate change, it is significant as the largest and most widely known and discussed report of its kind.

3 The IPCC Representative Concentration Pathway (RCP) scenario 8.5 assumes unabated emissions continue and shows the dangers of business as usual. RCP 8.5 refers to the concentration of carbon that delivers global warming at an average of 8.5 watts per square metre across the planet. RCP 8.5 delivers a temperature increase of about 4.3 degrees Celsius by 2100, relative to preindustrial temperatures. For more details, see https://climatenexus. org/climate-change-news/rcp-8-5-business-as-usual-or-a-worst-case-scenario.

References

Ackerman, J. (2020) *The Bird Way: A New Look at How Birds, Walk, Talk, Play, Parent and Think.* London: Penguin Press.

Alley, R. (2000) 'The Younger Dryas cold interval as viewed from central Greenland'. *Quarternary Science Reviews*, 19 (1): 213–226.

Bangkok Post. (2019) 'Experts warn of "endless" flood cycle', 21 September [Online]. Available at: www.bangkokpost.com/thailand/general/1755089/experts-warn-of-endless-flood-cycle (accessed: 12 August 2020).

BBC. (2015) 'When global warming made the world super-hot', 14 September [Online]. Available at: www.bbc.com/earth/story/20150914-when-global-warming-made-our-world-super-hot (accessed: 27 May 2021).

———. (2019) 'How to save a sinking island nation', 13 August [Online]. Available at: www.bbc.com/future/article/20190813-how-to-save-a-sinking-island-nation (accessed: 3 August 2020).

———. (2020a) 'Summers could become "too hot for humans"', 16 July [Online]. Available at: www.bbc.com/news/science-environment-53415298 (accessed: 17 July 2020).

———. (2020b). 'Arctic Circle sees "hottest-ever" recorded temperatures', 22 June [Online]. Available at: www.bbc.com/news/science-environment-53140069 (accessed: 22 June 2020).

———. (2020c) 'US West Coast fires: Row over climate change's role as Trump visits', 14 September [Online]. Available at: www.bbc.com/news/world-us-canada-54144651 (accessed: 14 September 2020).

Box, G.E.P. (1970) 'All models are wrong, but some are useful' [Online]. Available at: www.lacan.upc.edu/admoreWeb/2018/05/all-models-are-wrong-but-some-are-useful-george-e-p-box (accessed: 2 February 2021).

Boyle, M.J. 2020. 'Tragedy of the Commons', investopedia.com, 23 October [Online]. Available at: www.investopedia.com/terms/t/tragedy-of-the-commons.asp (accessed: 20 February 2021).

Brienen, R., Phillips, O., Feldpausch, T., *et al.* (2015) 'Long-term decline of the Amazon carbon sink'. *Nature*, 519: 344–348 [Online]. Available at: https://doi.org/10.1038/nature14283 (accessed: 1 June 2020).

Business Insider. (2019) 'Europe's mightiest river is drying up, most likely causing a recession in Germany. Yes, really', 22 January [Online]. Available at: www.businessinsider.com/germany-recession-river-rhine-running-dry-2019-1 (accessed: 12 August 2020).

Carney, M. (2015) 'Breaking the tragedy of the horizon – climate change and financial stability'. Speech, Youtube video, at mainstreamingclimate.org [Online]. Available at: www.mainstreamingclimate.org/publication/breaking-the-tragedy-of-the-horizon-climate-change-and-financial-stability (accessed: 29 January 2021).

Cook, K. and Vizy, E. (2007) 'Effects of 21st century climate change on the Amazon rainforest'. *Journal of Climate*, 21 (3): 542–560.

DeFries, R., Edenhofer, O., Halliday, A., Heal, G., Lenton, T. *et al.* (2019) 'The missing economic risks in assessments of climate change impacts'. Earth Institute, Policy Insight, September [Online]. Available at: www.lse.ac.uk/granthaminstitute/wp-content/uploads/2019/09/The-missing-economic-risks-in-assessments-of-climate-change-impacts-2.pdf (accessed: 20 February 2021).

Duncan, B. (2004) *The Wee Book of Calvin: Air Kissing in the North East.* London: Penguin.

Earth in the Future. (2021) 'Ancient clime event: The Paleocene Eocene thermal maximum' [Online]. Available at: www.e-education.psu.edu/earth103/node/639 (accessed: 7 February 2021).

Favier, L., Durand, G., Cornford, S.L., Gudmundsson, G.H., Gagliardini, O., Gillet-Chaulet, F., Zwinger, T., Payner, A.J., and Le Brocq, A.M. (2014) 'Retreat of Pine Island Glacier controlled by marine ice-sheet instability'. *Nature Climate Change*, 4 (12 January): 117–121 [Online]. Available at: www.nature.com/articles/nclimate2094 (accessed: 27 May 2020).

Fecht, S. (2020) 'Unusually clear skies drove record loss of Greenland ice in 2019.' *Columbia Earth Sciences*, 15 April [Online]. Available at: https://blogs.ei.columbia.edu/2020/04/15/clear-skies-greenland-ice-loss-2019/ (accessed: 27 May 2020).

Fleurbaey, M. and Zuber, S. (2013) 'Climate policies deserve a negative discount rate'. *Chicago Journal of International Law*, 13(2) [Online]. Available at: https://chicagounbound.uchicago.edu/cjil/vol13/iss2/14 (accessed: 7 February 2021).

Forbes. (2020) 'Warlike: Prince Charles calls tor "Marshall Plan" to fight climate change', 21 September [Online]. Available at: www.forbes.com/sites/davidrvetter/2020/09/21/warlike-prince-charles-calls-for-marshall-plan-to-fight-climate-change (accessed: 7 February 2021).

Fosler, G. (2013) 'What would Adam Smith say about morals and markets?', 14 January [Online]. Available at: www.gailfosler.com/what-would-adam-smith-say-about-morals-and-markets (accessed: 21 February 2021).

Friedman, M. (1970) 'The Social responsibility of business is to increase its profits'. *New York Times*, 13 September [Online]. Available at: http://umich.edu/~thecore/doc/Friedman.pdf (accessed 22 February 2021).

Geophysical Research Letters. (2020) 'Arctic sea ice in CMIP6', 17 April [Online]. Available at: https://agupubs.onlinelibrary.wiley.com/doi/full/10.1029/2019GL086749 (accessed: 22 May 2020).

Gladwell, M. (2000) *Tipping Point: How Little Things Can Make a Big Difference*. New York: Back Bay Books.

Graafland J. and Wells, T.R. (2019) 'In Adam Smith's own words: The role of virtues in the relationship between free market economies and societal flourishing, a semantic network data-mining approach'. *Journal of Business Ethics*. [Advance online]. Available at: https://doi.org/10.1007/s10551-020-04521-5 (accessed: 28 January 2021).

The Guardian. (2020a) 'Ice-free Arctic summers now very likely even with climate action' [Online]. Available at: www.theguardian.com/world/2020/apr/21/ice-free-arctic-summers-now-very-likely-even-with-climate-action (accessed: 22 May 2020).

———. (2020b) 'Climate worst-case scenarios may not go far enough', 13 June [Online]. Available at: www.theguardian.com/environment/2020/jun/13/climate-worst-case-scenarios-clouds-scientists-global-heating? (accessed: 15 June 2020).

Hardin, G. (1968) 'The tragedy of the commons: The population problem has no solution; it requires a fundamental extension in mortality'. *Science*, 162 (3859): 1243–1248.

Haq, B.U., Hardenbol, J., and Vail, P.R. (1987) 'Chronology of fluctuating sea levels since the Triassic'. *Science*, 235: 1156–1167 [Online]. Available at: https://science.sciencemag.org/content/235/4793/1156/tab-article-info (accessed: 20 February 2021).

Harper, K. (2017) *The Fate of Rome*. Princeton: Princeton University Press.

Harvey, C. (2020) '"Zombie" fires may be reigniting after Siberian winter'. E&E News, 2 June [Online]. Available at: www.eenews.net/stories/1063293625 (accessed: 28 January 2021).

Hayek. M. (1943) *The Road to Serfdom*. Chicago: University of Chicago Press. 2003 edition.

Hickel, J. (2016) "The Nobel Prize for climate catastrophe'. *Foreign Policy*, 6 December [Online]. Available at: https://foreignpolicy.com/2018/12/06/the-nobel-prize-for-climate-catastrophe/ (accessed: 28 January 2021).

Huss, M. (2017) 'Toward mountains without permanent snow and ice'. *Earth's Future*, 5: 418–435.

Inside Climate News. (2019) 'Losing Arctic ice and permafrost will cost trillions as Earth warms, study says', 23 April [Online]. Available at: https://insideclimatenews.org/news/23042019/arctic-permafrost-climate-change-costs-feedback-loop-ice-study (accessed: 22 May 2020).

Jougin, I., Smith, B.E., and Medley, B. (2014) 'Marine ice sheet collapse potentially under way for the Thwaites Glacier Basin, West Antarctica'. *Science*, 344 (6185): 735–738 [Online]. Available at: https://science.sciencemag.org/content/344/6185/735 (accessed: 27 May 2020).

Keen, S. (2019) 'The cost of climate change.' *Evoeconomics* [Online]. Available at: https://evonomics.com/steve-keen-nordhaus-climate-change-economics (accessed: 7 February 2021).

Krugman, P. (2020) 'Apocalypse becomes the new normal', *New York Times*, 2 January [Online]. Available at: www.nytimes.com/2020/01/02/opinion/climate-change-australia.html (accessed: 14 August 2020).

Lapola, D.M., Pinho, P., Quesada, C.A., Bernardo, B.N., Strassburg, A., Rammig, B.K., *et al.* (2018) 'Limiting the high impacts of Amazon forest dieback with no-regrets science and policy action'. *Proceedings of the National Academy of Science*, 115 (46): 11671–11679 [Online]. Available at: www.pnas.org/content/115/46/11671 (accessed: 21 May 2020).

Lenton, T., Held, H., Kriegler, E., Hall, J.W., Lucht, W., Rahnstorf, S., and Schellnhuber, H.J. (2008) 'Tipping elements in Earth's climate system'. *Proceedings of the National Academy of Sciences of the United States of America*, 12 February [Online]. Available at: www.pnas.org/content/105/6/1786 (accessed: 28 January 2021).

Lustgarten, A. (2020). 'The great climate migration', *New York Times*, 23 July [Online]. Available at: www.nytimes.com/interactive/2020/07/23/magazine/climate-migration.html?referringSource=articleShare (accessed: 3 August 2020).

Marzeion, B., Cogley, J.G., Richter, K., and Parkes, D. (2014) 'Glaciers: Attribution of global glacier mass loss to anthropogenic and natural causes'. *Science*, 345 (6199): 919–921. doi: 10.1126/science.1254702.

McKinsey & Company. (2020) 'Climate risk and response: Physical hazards and socio-economic impacts', 16 January [Online]. Available at: www.mckinsey.com/business-functions/sustainability/our-insights/climate-risk-and-response-physical-hazards-and-socioeconomic-impacts (accessed: 28 January 2021).

Monty Python. (1979) The Life of Brian [Online]. Available at: www.youtube.com/watch?v=WoaktW-Lu38 (accessed: 19 January 2021).

NASA (National Air and Space Administration). (2019). 'Huge wildfires in Russia's Siberian province continue', 16 August [Online]. Available at: www.nasa.gov/image-feature/goddard/2019/huge-wildfires-in-russias-siberian-province-continue (accessed: 28 May 2020).

Neuman, A.E. (1954) The first appearance of Mr. Neuman on the cover of The Mad Reader. Sourced from: Reidelbach, M. (1992). *Completely Mad: A History of the Comic Book and Magazine*. New York: Little Brown & Company.

New York Times. (2018) 'Trump put a low cost on carbon emissions: Here's why it matters', 23 August [Online]. Available at: www.nytimes.com/2018/08/23/climate/social-cost-carbon.html (accessed: 22 January 2021).

———. (2020) 'Rising seas threaten an American institution: The 30-year mortgage', 19 June [Online]. Available at: www.nytimes.com/2020/06/19/climate/climate-seas-30-year-mortgage.html?referringSource=articleShare (accessed: 19 June 2020).

NOAA (National Oceanic and Atmospheric Administration). (2019) 'Hurricane fast facts' [Online]. Available at: https://coast.noaa.gov/states/fast-facts/hurricane-costs.html (accessed: 10 June 2020).

———. (2020) 'The Younger Dryas' [Online]. Available at: www.ncdc.noaa.gov/abrupt-climate-change/The%20Younger%20Dryas (accessed: 21 May 2020).

Nordhaus, W.D. (2017) 'Revisiting the social cost of carbon'. *Proceedings of the National Academy of Sciences of the United States (PNAS)*, 114 (7): 1518–1523, 14 February [Online]. Available at: www.pnas.org/content/114/7/1518 (accessed: 9 February 2021).

———. (2018) 'Climate change: the ultimate challenge for economics'. Nobel Prize Lecture [Online]. Available at: www.nobelprize.org/prizes/economic-sciences/2018/nordhaus/lecture (accessed: 7 February 2021).

NSIDC (National Snow and Ice Data Center). (2019) 'Large ice loss on Greenland Ice sheet in 2019', 8 November [Online]. Available at: http://nsidc.org/greenland-today/2019/ (accessed: 27 May 2020).

Olsson, R. (2009) 'Boreal forest dieback may cause runaway global warming'. Fact Sheet No. 22, Air Pollution and Climate Secretariat, Göteborg, Sweden.

Parramore, L. (2019) 'Are economists blocking progress on climate change?' Institute for New Economic Thinking, 24 June [Online] Available at: www.ineteconomics.org/perspectives/blog/are-economists-blocking-progress-on-climate-change (accessed: 28 January 2021).

Pelto, M. (2019) 'Alpine glaciers: Another decade of loss'. *RealClimate*, 25 March [Online]. Available at: www.realclimate.org/index.php/archives/2019/03/alpine-glaciers-another-decade-of-loss (accessed: 28 May 2020).

Pindyck, R.S. (2017) 'The use and abuse of models for climate policy'. *Review of Environmental Economics and Policy*, 11 (1): 100–114.

Pinker, S. (2012) *The Better Angels of Our Nature*. New York: Penguin Books.

Rand, A. (1957) *Atlas Shrugged*. Mass market paperback. 1996 edition.

Rasmussen, D. (2017) *The Infidel and the Professor: David Hume and Adam Smith, and the Friendship that Shaped the Modern World*. Princeton: Princeton University Press.

Sagar, P. (2018) 'Adam Smith and the conspiracy of the merchants'. *Global Intellectual History* [Online]. Available at: doi:10.1080/23801883.2018.1530066 (accessed: 28 January 2021).

Schoof, C. (2007) 'Ice sheet grounding line dynamics: Steady states, stability, and hysteresis'. *Journal of Geophysical Research*, 112: 1–19 [Online]. Available at: https://agupubs.onlinelibrary.wiley.com/doi/pdf/10.1029/2006JF000664 (accessed: 27 May 2020).

Shiller, R. (2019). *Narrative Economics: How Stories Go Viral and Drive Major Economic Events*. Princeton: Princeton University Press.

Smith, A. (1776) *An Inquiry into the Nature and Causes of the Wealth of Nations*. New York: Bantam Classics. 2003 edition.

———. (1759) *The Theory of Moral Sentiments*. Project Gutenberg [Online]. Available at: www.gutenberg.org/files/58559/58559-h/58559-h.htm (accessed: 20 February 2021).

Steffen, W., Rockström, J., Richardson, K., Lenton, T.M., Folke, C. *et al.* (2018) 'Trajectories of the Earth system in the Anthropocene'. *Proceedings of the National Academy of Sciences of the United States*, 115 (33): 8252–8259 [Online]. Available at: http://hdl.handle.net/2078.1/204292 (accessed: 20 February 2021).

Stern, Nicholas. (2006) *The Stern Review on the Economics of Climate Change*. Government of the United Kingdom, London, 27 November [Online]. Available at: https://onlinelibrary.wiley.com/doi/abs/10.1111/j.1728-4457.2006.00153.x (accessed: 20 February 2021).

Stiglitz, J. (2019) 'Critics of the Green New Deal ask if we can afford it. But we can't afford not to: Our civilisation is at stake', *The Guardian*, 4 June [Online]. Available at: www.theguardian.com/commentisfree/2019/jun/04/climate-change-world-war-iii-green-new-deal (accessed: 7 February 2021).

Swiss Re. (2020) 'A fifth of countries worldwide at risk from ecosystem collapse as biodiversity declines, reveals pioneering Swiss Re index', 23 September [Online]. Available at: www.swissre.com/media/news-releases/nr-20200923-biodiversity-and-ecosystems-services.html (accessed: 17 May 2021).

Taleb, N. (2010) *The Black Swan: The Impact of the Highly Improbable.* Second edition. New York: Random House.

Thames Barrier. (2020) 'The Thames Barrier' [Online]. Available at: www.gov.uk/guidance/the-thames-barrier (accessed: 28 January 2021).

Thunberg, Greta. (2019) Speech delivered at the United Nations Climate Action Summit, New York, 23 September [Online]. Available at: www.youtube.com/watch?v=KAJsdgTPJpU (accessed: 7 February 2021).

Troianovski, A. and Mooney, C. (2019) 'Radical warming in Siberia leaves millions on unstable ground'. *National Geographic*, 3 October [Online]. Available at: www.washingtonpost.com/graphics/2019/national/climate-environment/climate-change-siberia/ (accessed: 1 June 2020).

University of Sydney. (2020) 'A statement about the 480 million animals killed in NSW bushfires since September', 3 January [Online]. Available at: www.sydney.edu.au/news-opinion/news/2020/01/03/a-statement-about-the-480-million-animals-killed-in-nsw-bushfire.html# (accessed: 18 June 2020).

USDA (United States Department of Agriculture). (2020) 'The cost of raising a child', 18 February [Online]. Available at: www.usda.gov/media/blog/2017/01/13/cost-raising-child (accessed: 28 January 2021).

Wadhams, P. (2016) 'The global impact of rapidly disappearing arctic sea ice'. YaleEnvironment360, 26 September [Online]. Available at: https://e360.yale.edu/features/as_arctic_ocean_ice_disappears_global_climate_impacts_intensify_wadhams (accessed: 28 January 2021).

Wagner, G. and Weitzman, M. (2015) *Climate Shock: The Economic Consequences of a Hotter Planet.* Princeton: Princeton University Press.

Ward, B. (2018) 'Climate economics is based on models that are not fit for purpose', *Financial Times*, 28 December [Online]. Available at: www.ft.com/content/a486e482-09d8-11e9-9fe8-acdb36967cfc (accessed: 28 January 2021).

Washington Post. (2020) 'Baghdad soars to 125 blistering degrees, its highest temperature on record' [Online]. Available at: www.washingtonpost.com/weather/2020/07/29/baghdad-iraq-heat-record/?fbclid=IwAR3cY3hC7eyU5JqQv5In8uXzuzxds5gmgdUkSAvdnIKGy6dwSG80Z2AN1xM (accessed: 1 August 2020).

Weitzman, M.L. (2011) 'Fat-tailed uncertainty in the economics of catastrophic climate change'. *Review of Environmental Economics and Policy*, 5 (2):275–292.

World Bank. (2018) 'Climate change could force over 140 million to migrate within countries by 2050' [Online]. Available at: www.worldbank.org/en/news/press-release/2018/03/19/climate-change-could-force-over-140-million-to-migrate-within-countries-by-2050-world-bank-report (accessed: 18 June 2020).

Wright, J.D. and Schaller, M.F. (2013) 'Evidence for a rapid release of carbon at the Paleocene-Eocene thermal maximum'. *Proceedings of the National Academy of Sciences of the United States*, 110 (40): 15908–15913, 1 October [Online]. Available at: https://doi.org/10.1073/pnas.1309188110 (accessed: 20 February 2021).

Wucker, M. (2016) *The Gray Rhino; How to Recognize and React to the Obvious Dangers We Ignore.* New York: St Martin's Press.

Yumashev, D., Hope, C., Schaefer, K., *et al.* (2019) 'Climate policy implications of nonlinear decline of Arctic land permafrost and other cryosphere elements'. *Nature Communications*, 10 (1900) [Online]. Available at: www.nature.com/articles/s41467-019-09863-x (accessed: 22 May 2020).

Zemp, M., Frey, H., Gartner-Roer *et al.* (2015) 'Historically unprecedented global glacier decline in the early 21st century'. *Journal of Glaciology*, 61 (228): 745–762.

3

SETTING TARGETS, PRICING CARBON, AND PUNISHING LAGGARDS

> Economists have had a long predilection for price interventions to correct market failures such as those arising from the presence of externalities. The reason is simple: market efficiency requires equating private and social returns, the presence of an externality means that there is a gap between the two, and a price intervention can close the gap, restoring efficiency.
>
> *Stiglitz, 2019*

> I do think if [carbon] risk is priced, it will drive the right behaviors ... this is an economy wide transition where every asset will see a change in value.
>
> *Breeden, 2021*

Securing net zero begins with ambitious stretch commitments to achieve our goal by 2050 and secure a stable and relatively temperate climate. To date, over 120 countries have committed to the net-zero goal but getting from polluting today to net zero by 2050 will require a fivefold acceleration in GHG emission reductions. Publicly stated commitments are only a first step. Commitments need to be followed up with consistent implementation, monitoring, goal adjustment, and enforcement across all sectors, affecting all parts of our societies and economies. Now is not the time for incremental change or to falsely reassure people that getting to net zero requires little change in the way we work and operate. Significant commitments must be backed by broad, decades-long governmental and regulatory action. Making commitments is essential, but this is only the (often belated) first step.

Setting ambitious net-zero targets should be matched by a global agreement to price carbon with progressive increases over time and so affect incentives, markets, and individual behaviours. Doing so is an economic necessity (Stiglitz, 2019). Pricing carbon will begin the process of shifting incentives (Breeden, 2021). A clear

DOI: 10.4324/9781003037088-4

policy consensus is coalescing that pricing carbon is required to accelerate us to our goal. In 2021, scores of countries price carbon through taxation and cap-and-trade schemes. However, short-term political pressures, demands, and resistance continue to trump planetary needs, and as a result carbon prices are invariably set far too low to achieve desired goals.

Negotiators preparing for COP26 know carbon taxes and mechanisms exist in many markets (though not yet in the US), but the levels and rates are insufficient to achieve net-zero goals. Leaders need to make clear that this has to change, and fast. Governments and leaders must add momentum to the narrative, policy, and market shift that is beginning by making meaningful commitments and raising carbon prices globally and nationally. A high minimum and a commitment to rising prices needs to be announced and laid out clearly, market to market, across all countries. Doing so will accelerate dynamics already underway, harness markets, pull forward investment decisions, shift business strategies, reward the innovative, and punish the polluters. The process of setting targets, pricing carbon sufficiently highly, and securing GHG reductions will, however, be politically contentious, but it should be manageable. Once in place, carbon pricing would help bend the curve on GHG emissions as well as spur decades of sustainable growth and help solve our productivity and secular stagnation conundrums.

Governments that act and address net zero through rising carbon prices should proceed regardless of whether there are those (and there will be some) that refuse to act and pursue net-zero goals. The states and actors that refuse to pay for carbon pollution should pay a real political, diplomatic, and economic price for their selfish behaviour. In pricing carbon effectively, states cannot permit laggards and free riders to abuse the global commons any further.

Leaders of countries actively and consistently committed to net-zero goals should form a coalition of the willing and construct a digitized, renewably electrified Green Globalization 2.0. This regreened globalization must champion free trade, but this trade must be based on the real cost of carbon and planetary burdens that are required to secure net zero. We need enforcement mechanisms to ensure that Green Globalization 2.0 is sustainable, resilient, and bends the market and our economies to the planetary GHG and societal goals. The days of championing unrestricted trade rights of planetary polluters who refuse to pay the real economic cost of their dangerous carbon addiction must end.

Getting to net zero will require that Green Globalization 2.0 is supported and protected via tariffs against the states, firms, and actors that fail to price carbon at the agreed rising level. The world can construct a Green Globalization 2.0 that rejects unfettered access to markets by those that pollute the global commons.

This chapter addresses the importance of commitments, rising carbon prices, the creation of a coalition of the willing, and enforcement of our common carbon-price goals. As with so much else in the climate change space, we know what is necessary. We have examples that show us what works well, what works less well, and how we can proceed without unanimity. We also know that enforcement encourages and supports fair trade and economic growth. Properly constructed

Green Globalization 2.0 can be a decarbonized reindustrialization for the twenty-first century. A renewed and reseeded growth based on 50 shades of green is the solution.

Let us turn first to the matter of ambitious commitments and targets, for without making a commitment there is procrastination and delay, neither of which we can afford at this climate juncture.

Set ambitious long-term targets and match them with clear interim steps and plans

Leaders preparing for COP26 should set increasingly ambitious long-term targets and back them with clear, publicly declared interim steps and plans. In 2021, more than 120 countries have paid lip service to net-zero goals. If humanity is to have a real chance of reaching net zero by 2050, the leading polluting nations and regions of the world need to drastically increase the scale of their ambitions in terms of GHG emissions reduction goals. Sadly, in the past, policymakers' and states' climate change IPCC commitments have too often proven to be woefully inadequate in ambition, and states have failed to deliver GHG reductions. Sceptics will note that we have seen this all before, at COP meetings in the past *ad nauseam*, in Kyoto, in Rio, even in Paris. Leaders gathered and IPCC scientists raised their collective voices in increasing alarm. Nonbinding and unenforced agreements were usually announced, yet most states fail to live up to their commitments. Can it be different in 2021? Can the cycle of under-promising and under-delivering be broken once and for all?

The run-up to COP26 negotiations provides a key test and an opportunity to change the dynamic from disappointment and delay to one of anticipation and action. Governments should set more aggressive goals for their own GHG paths to net zero. As UN Secretary-General António Guterres has stated: 'If we are going to limit global heating to 1.5 degrees Celsius, we need to demonstrate, starting this year, how we will achieve emissions reductions of 45 percent from 2010 levels this decade, and how we will reach net-zero emissions by mid-century' (UN, 2020).

At COP26, states must underline the seriousness with which they view the challenge. States can then marshal their governments, their spending, and their state power to the task. States can progressively align goals with policy practice and implementation, and embed 2050 environmental goals, reform, and renewal within all their short-, medium-, and long-term organizational plans and objectives. States must internalize and integrate climate crisis goals in every aspect of government planning.

This internalizing of net-zero goals, which is already underway in some states, societies, and sectors, must become a standard test for ethical, acceptable, sustainable good governance in our parliaments and businesses, and in economics.

All should ask: 'Does policy A, business choice B, long-term strategy C fit with our net-zero goals and alignment?' If the answer is no, the choice should be rejected. If a bad option is selected regardless, the carbon cost, penalties, and disincentives must be large and become progressively greater the further an approach is from agreed net-zero targets. Polluters must pay an increasingly high price to spew GHG into the atmosphere.

New, more ambitious targets can help shift expectations among communities, businesses, and investors and begin to build expectations of relatively rapid transformative change. Ambitious goals, once established, create forces and dynamics that work in favour of success; they can herald the start of our war on carbon (BBC, 2020). States, governments, regions, and cities can then be repeatedly measured against their goals by activists, technocratic observers, their increasingly alarmed voting public, and Greta and her young school strikers for climate. States and our leaders must be held to account and be forced to measure success against their own net-zero commitments. Remarkably, in 2020 and 2021, it appears key leaders and states have begun to extend and enlarge their climate change goals and the scale of their ambitions.

Europe's regreening and rebuilding

After the Paris Agreement, the EU set an ambitious goal – 40 percent GHG reduction by 2030. This was subsequently revised and increased in the European Green Deal, which was signed in December 2019, in which the EU announced a commitment to net zero by 2050 and a commitment to a new GHG reduction goal of 55 percent by 2030.

The European Green Deal will embed the 2050 GHG reduction goals into EU law and establish a consultative process to engage citizens in this collective endeavour. The European Green Deal seeks to:

- Set the long-term direction of travel for meeting the 2050 climate-neutrality objective through all policies
- Create a system for monitoring progress and take further action as required
- Provide predictability for investors and other economic actors
- Ensure that the transition to climate neutrality is irreversible.

This will be a mammoth task involving 'All parts of society and economic sectors … from the power sector to industry, mobility, buildings, agriculture and forestry' (EU Commission, 2018). It was, according to EU Commission President Ursula von der Leyen, the continent's 'man on the Moon' moment (BBC, 2019). The effort will require a major shift in the structure of the European economy towards greater reliance on renewables and energy efficiency, among much else.

EU leaders signalled their seriousness to financial markets with a simultaneous commitment by the European Investment Bank (EIB) – the EU's lending arm – that it would provide 1 trillion euros of investment in climate action and environmental sustainability from 2021 to 2030.

Since this key step, European policymakers have explicitly linked the commission's 750-billion-euro Next Generation fund to building the green economy of tomorrow and addressing the climate emergency and Covid-19 recovery.

The European Green Deal leverages a further 1 trillion euros in public and private funds for green investment, more stringent standards for investment designed to avoid greenwashing, 100 billion euros towards programmes for a just transition, and the designing of a new industrial policy including elements such as a strategy for a circular economy[1] and a sustainable food system.

Just as Europe's fiscal authorities are reaching for stretch targets, so too is Christine Lagarde, president of the European Central Bank (ECB). Lagarde is pursing the integration of climate change goals and risks within the ECB mandate and policy actions. She sees this as an essential and urgent goal for the most powerful central bank on the planet. The ECB is signalling to markets and investors that it is throwing its weight behind Europe's decarbonization goal, that change is coming, and that this lender of last resort understands climate risks and the need for action.

For Europe, the rate of the managed transition ahead – the glidepath – will be steep (see Figure 3.1). Yet surely a managed decent is better than a precipitous plummet. Overall, the EU is ahead of the rest of the world in GHG reductions. Many European states, including Finland, France, Scotland, Sweden, and the UK, are bending the curve on GHG emissions and the rate of transition. The key to success is found in the shifting public and policy narrative, the numerous waypoints, the national and European incentives, and the enforcement mechanisms and monitoring that takes policy ambitions and turns them into a reality in the air and on the ground. Measured against goals set in 2008 (not the current, more aggressive, goals), the EU has performed well. Its leaders have – rightly – increased their ambition progressively. Figure 3.1 shows a GHG reduction of 23.2 percent over 1990

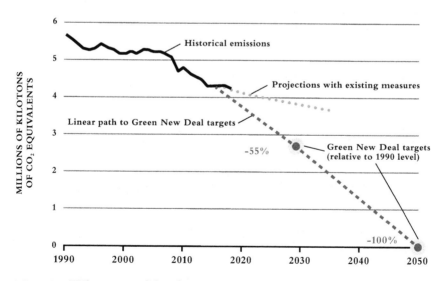

FIGURE 3.1 EU's net-zero glidepath

Source: United Nations Framework Convention on Climate Change; European Environment Agency.

levels (EU Commission, 2020). The new reduction goals are more aggressive and are needed to get to the 2050 net-zero goal.

In 2021, the EU states (and the UK) are some of the more advanced in their Paris Agreement goals and glidepaths. Some states, such as France, Italy, Sweden, and the UK, have legally binding commitments to net zero and falling per capita GHG emissions because of policy actions and responses. Still, hitting targets for GHG emissions is proving difficult. It is being tackled by closing power stations (in France and the UK) and with altered incentives and penalties. In other EU countries, such as Belgium, Denmark, Germany, Ireland, and the Netherlands, governments are failing to deliver on agreed goals. They must up their game.

The EU, despite its many challenges, has effective political and policymaking processes on climate change goals and the glidepath. It is setting targets, progressively increasing them, setting up monitoring processes, and applying enforcement measures, in this case through well-understood existing mechanisms and the European legal system.

Pursued consistently, this could redirect resources and reorient and regreen the continent's industrial policies. All these steps, supported fiscally via regulatory changes and on the monetary side by the ECB shift, will speed the transition and increase the possibility of Europe achieving the goal and grasping the economic upside of an industrial shift that is already underway but for which many further policy, pricing, and incentive steps are required.

A Chinese commitment with potentially momentous impact

In a potentially game-changing move, the Chinese government raised its level of ambition in 2020, with an announcement it would reach net zero by 2060. This is the first time China, which emits 28 percent of global GHG, has made an explicit timeline commitment to net zero. President Xi Jinping's move 'took the West by surprise … climate change politics at a global level shifted into a new gear' (Tooze, 2020a). If pursued consistently, this is a major alteration in the global GHG and climate calculus.

China committed to reach a peak in GHG emissions in 2030 and decline from there to net zero by 2060. The details of the plan remain obscure, but the ambition is significant and the shift, once it is backed by China's planning might, significantly alters the global and COP26 diplomatic, political, and economic struggle from one that seemed increasingly impossible to one that may be possible given a successful COP26 and national follow-through and progressive implementation. China's move immediately catapults it into a leadership position on climate change. As Tooze (2020a) notes:

> China has now doubled down on the Paris framework. For the first time since climate talks began in the early 1990s, the largest emitter has committed to decarbonization.

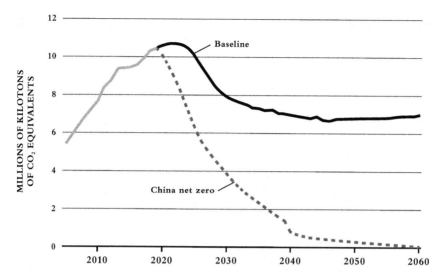

FIGURE 3.2 China's net-zero glidepath

Source: United Nations Framework Convention on Climate Change; Carbon Brief. See www.carbonbrief.org/analysis-going-carbon-neutral-by-2060-will-make-china-richer.

China's future GHG reductions will have to be very steep indeed, across all aspects of its economy (see Figure 3.2).

Unlike the EU, which has been transitioning away from GHGs for years, China is the most carbon-intensive economy in the industrialized world. As the International Energy Agency (2020) notes, China has an

> Energy-intensive growth model and a carbon-intensive energy supply [which has] created an enormous carbon footprint. In the last 20 years, CO_2 emissions in China grew six times as fast as in the rest of the world, and China accounted for almost two-thirds of the growth in global CO_2 emissions. … the statement by President Xi Jinping … that China would strive to be carbon neutral by 2060 should not be seen as a technical change in the details of energy and environmental policy. Rather, it potentially represents the biggest climate undertaking ever made by any country.

The crucial test for President Xi will be contained in the contents of the next five-year plan. If China follows its net-zero commitment with detailed short-, medium-, and long-term implementation plans, enforcement mechanisms, and monitoring systems, the commitment can rapidly impact and speed industrial transformation. Done right, it would not only create a new, low-carbon energy system but also a new economy and society in China, as well as help secure a sustainable, resilient, decarbonized economy. But can it be done?

Those who doubt whether China can go from polluting giant today to a net-zero 2060 should not underestimate the Chinese capability for extremely rapid industrial and technological change. The Chinese state can achieve rates of industrialization, transformation, and radical jumps often as fast, or faster, than other nations.

For instance, in the heat of the building and economic boom of the 2000s, China poured more cement in one five-year period than was used in the US in the entire twentieth century. In another instance of massive industrial planning and might, China built a 150,000-kilometre interstate road network, the largest in the world, by 2019, having begun only 35 years earlier. China also constructed the world's largest high-speed electric rail network – some 35,000-kilometres long – in 14 years. Contrast this with the US, which in 2021 does not have a single high-speed line in the whole country. Or consider the UK, which takes decades to construct a single, short, high-speed line after endless arguments and recriminations.

If there is one thing China knows how to do, it is to plan for sudden change and execute it rapidly using all the levers of state authority and power. That is what they will do in this, the greatest challenge facing the Chinese state and their 'communism with Chinese characteristics' model. We should all be hoping they pull this industrial transformation off, for they must do so if we are to achieve our common net-zero goals. The greening of China has just started. I expect most of us will be surprised by the pace of change once the power of the Chinese state is fully engaged. So, Europe and China are making stretch goals public and signalling a shift. What about America?

Biden's Green New Deal – a start, finally

America's absence from the climate change debate, the government's active hostility to climate change science, its exit from the Paris Agreement, and its rollback of US environmental standards from 2016 to 2021 directly threatened the globe's climate and net-zero goals. The election of Joseph R. Biden to the presidency signals a return of US leadership and a reengagement of US diplomacy towards the common goal of net zero. The new administration's clear commitment to climate change as a crisis and strategic area for coordinated action is extremely important for the world.

As a candidate in 2020, Biden announced his support for a US$1.7 trillion investment plan – the Green New Deal (GND). He stressed that science tells us we have nine years before the damage is irreversible and committed to getting the US to net zero by 2050. Biden's recognition that the situation is urgent, and his commitment to a massive investment plan to achieve the carbon goal, meaningfully alters industry and investor plans. His endorsement of science rather than garbage conspiracies and denial is a refreshing and urgently needed return to policymaking based on facts, not falsehoods. The US has a long way to go to meet its goals, and the steepness of its transition will be quite marked given the failure of the Trump and other past administrations (including, sadly, that of President Obama) to bend the GHG curve meaningfully.

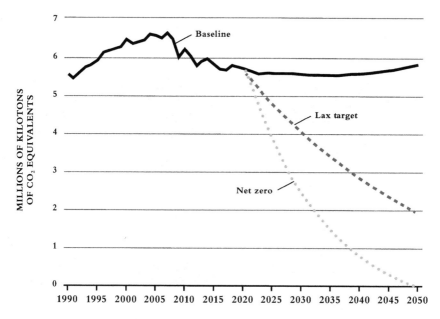

FIGURE 3.3 United States' net–zero glidepath

Source: American Action Forum, 2021.

Leadership at last

Leadership always matters. There has been a meaningful and striking shift in the rhetoric on climate change, and President Biden is embedding climate change policy all across government. It is no coincidence that every single cabinet-level official cites the urgency of climate change as a principal policy goal. Rather, it reflects a carefully thought-out policy narrative driven from the very top. President Biden understands the importance of this shift – of the urgency and the danger inaction poses to the American economy. In reorienting the public policy narrative, he brings the US administration back into line with a plurality of US voters, who are increasingly alarmed by climate change and the risks it poses. President Biden is also politically astute. He knows who delivered his victory – predominantly younger people and students – generations X and Y – and educated, urban, and suburban voters – who rank climate change risks high on their list of concerns. In a sense, President Biden is not taking a political risk in reorienting the public policy story on climate change towards solutions and action; rather, he is agreeing with his supporters and the public.

President Biden's actions – his rejoining the Paris Agreement; his commitment to net zero; his hosting of a major summit in the spring of 2021; his reengagement with the COP process; his unveiling of a new green industrial policy, a policy leap which might be said to 'Make America Green Again' – permit America and Americans to put aside misinformation and conspiracy theory and turn towards the tasks ahead. However, commitments are one thing; implementation is the real test. California, America's largest and most prosperous state, has been showing the way forward.

California's demonstration of what is possible with leadership

California has already reached its own climate physical, narrative, and policy tipping points. California's record-breaking 2020 wildfires showed us in real time what a rapidly warming planet looks like. The future is here. The blazes tore through Trumpian distortion, lies, and misinformation. Voters' views in California reflect this. What is the single most important issue for Californians today? Climate change. In California, few deny the climate change reality or dispute the need to act, for when you cannot go outside, when you smell the smoke and see the fires, you see the reality.

California's narrative consensus has allowed Governor Gavin Newsom to respond to the fires by announcing a complete phase-out of petrol combustion engines by 2045 with no appreciable voter backlash. Automakers are now rushing to respond so as to not lose market share. California's move hastens electric vehicle (EV) adoption in America, as the largest market dictates investment plans for smaller ones. This single step by California will spur innovation, further reward first movers (witness US EV equities rising and rising), and penalize polluters accordingly. California's policy shifts will hasten creative destruction and herald the creation of scores of new firms and new employment opportunities. Similar dynamics in other countries, from Japan to Costa Rica to South Africa, are playing out. Countries are delineating not only their net-zero commitments but also their national glidepaths, wayposts, national monitoring, enforcement mechanisms, and reporting requirements.

What of the laggard states?

The net-zero shifts in the EU, China, and the US are putting growing pressure on the laggards such as Brazil, India, and Russia, and on other populist, nativist, nationalist polluters to increase their commitments and timelines lest they get left behind and well outside the rapidly evolving global consensus on climate crisis policy. This is precisely the type of dynamic we need in the run-up to COP26 and beyond – growing pressure for aggressive commitments followed by implementation plans. Just as Covid-19 hastened digital and other shifts in the US and global economy, so too can these significant and meaningful climate change policy announcements and their implementation speed that transformation.

A crucial element in securing our GHG goals requires the internalization of the cost of carbon – that is, we must price carbon to reflect its polluting and destructive effects and in doing so speed the rate of transition, the shift of economic incentives, and the behaviour of markets, investors, and individuals.

At COP26, governments must make the leap on enforceable and rising carbon pricing

Economists have long recognized that societies should price carbon appropriately and use markets and price mechanisms to shift incentives and speed the rate of change, cut GHG emissions, and help achieve climate goals. The simplest option in 'climate crisis economics 101' is a carbon price or tax, the setting of a transparent and rising cost on carbon that internalizes what has hitherto been an externality and in doing so triggering markets, firms, investors, and consumers to change their behaviours, choices, and lifestyles. A carbon price is comparatively easy to explain and understand. Voters, investors, and markets can adjust for the price, factor in cost increases, and make judgements accordingly.

Pricing can trigger rapid changes in behaviours

A carbon price can be applied as a direct tax, the simplest and most efficient option. It can also be applied within a cap-and-trade system. To achieve the Paris Agreement goals of net zero by 2050, most governments will use a combination of direct tax and cap and trade (as well as others), since neither is currently high enough nor covers enough of the economy's sectors to achieve the necessary climate change GHG goals.

The use of multiple tax and charge levers should be no surprise and is in line with normal policy approaches. Thus, we should increase petrol taxes to discourage driving; perhaps also increase the tax on SUVs to discourage the use of these vehicles; charge a congestion tax to discourage driving into cities; and apply a utility carbon tax or airline tax, or run an ETS programme for utilities to further change incentives. The precise balance and mix of policies depend on the political economy dynamics and demands in society.

At COP26, states should support the updating and redesign of national and regional carbon pricing (through taxation and cap-and-trade schemes) and help speed and solidify the shift in incentives. Time is short, though. At present, carbon pricing remains woefully inadequate globally.

Time to announce a 'C-Day'

Leaders coming together in Glasgow in 2021 should announce a global 'C-Day' – that is, a Commitment Day – and commit to a realistic minimum and progressively rising carbon price, country to country across the globe, with all states agreeing to levy a price on carbon by 2023. The time for vague commitments to pricing must end. Leaders and governments should signal that they all will put a price on carbon and commence the war on carbon. Government leaders should, where possible, strive for universal goals applied nationally and regionally across the planet. A minimum carbon price should be agreed and announced, and governments should set a series of targets for progressively rising carbon prices and a glidepath for pricing going forward, converging at a common high level in the medium to long term, and should commit to the creation of the mechanisms and waymarkers they will use to implement the pricing system. It will be up to nations to implement commitments buttressed by international coordination, national enforcement, and national monitoring mechanisms (these enforcement aspects are addressed in Chapter 5).

If at COP26 leaders confirm a collective commitment to an economically meaningful, inclusive, multisector, increasing carbon price in the three decades running up to 2050, businesses can adjust their strategies and business plans and invest accordingly. If carbon is priced effectively and consistently, investors will adjust. Consumers will respond to the signals they receive, investing here, consuming less there, disinvesting from firms that disregard their social and environmental goals. Business leaders are clear that what they need most of all to plan for a manageable climate change glidepath to net zero are the following regulatory factors.

The price of carbon – set a minimum, a timeline, and an adjustable glidepath

• *Establish a minimum carbon price and a line of sight in terms of a glidepath.* Governments and regulators need to establish a minimum price of carbon and the route to reaching that goal. These are starting to take shape through the commitments from China and the EU.

• *Establish a timeline during which a series of clear interim goals must be achieved.*

• *Provide a degree of policy certainty.* Businesses need to have some reassurance that policy will take X path, over Y timeline, with approximately Z price implications. With this knowledge, business leaders can adjust strategy. COP26 leaders must provide enhanced clarity about process and directional markers and waystations.

• *Provide clarity on metrics and data.* Businesspeople need to know when there is uncertainty and what factors might influence future policy shifts within an agreed framework. Economists constantly strive to understand a central bank's forward guidance and key data points influencing decisions and interest rate shifts. Businesspeople need similar clarity about directional markers for climate change.

States should set minimum carbon prices, the directional upward trajectory, markers, and potential steepness, as well as data points that markets and societies can work towards. National and regional authorities can build and strengthen market oversight, enforcement reforms, and planning processes. Nations and communities can pursue long-term investment and industrial policies designed to support the carbon price shift and underpin private sector investment with public infrastructure additions and goals. Communities can continue to plan for a net-zero and carbon-neutral future, which need not be some dystopian nightmare.

Regardless of the precise price, in advance of a global C-Day, investors will continue to move away from polluters' equity and debt. The risk premium demanded by investors will rise, especially for companies that fail to explain and demonstrate a path to zero. Polluters of course may retain the option of belching carbon now and closure tomorrow, but most would not. The market signals and mechanisms will work in our collective favour and the transition will be brought forward.

The path to agreement on, and the parameters of, a C-Day, is far from assured and is politically difficult, even if economically manageable if it is turbocharged in 2021. Nonetheless, government leaders must stretch to set the parameters of raising national carbon prices. Politicians should leave the implementation to other national and international delegated authorities, who are better able to oversee such markets and structures.

Setting a carbon price, however it is levied, must be based on a minimum price with an expectation and understanding that the price will gradually rise over time as per Figure 3.4. This approach is backed by the Carbon Leadership Council of central bankers and economists, such as current US Treasury Secretary Janet Yellen; Larry Summers, Treasury Secretary under the Obama administration; and many others. As Yellen states: 'When the central problem is the damage caused by greenhouse gas emissions, the cleanest and most efficient way to address it is to tax those emissions' (*Financial Times*, 2018). The Carbon Leadership Council has proposed setting a US$40 tax on each ton of CO_2 emitted in the US, with the price rising over time. The Carbon Leadership Council proposal is revenue neutral. All the revenue generated by the tax would be passed back to Americans as an annual dividend, thus ensuring equity and lessening or entirely alleviating the economic cost for most families.

Many countries have begun to tax carbon, albeit at too low a level, as Figure 3.5 shows. Sweden leads the world with a tax of US$130, followed by Liechtenstein

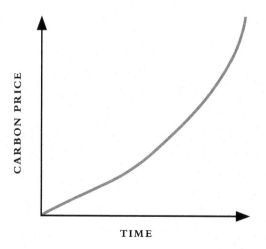

FIGURE 3.4 Setting a gradually rising carbon price to shift incentives

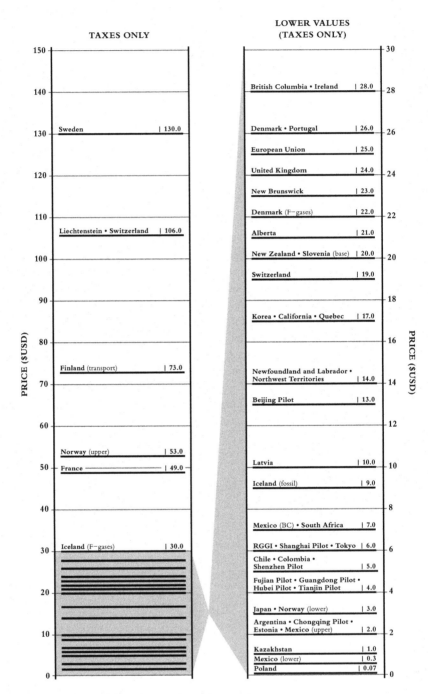

FIGURE 3.5 The world begins to price carbon

Source: World Bank, 2020.

(US$106), Finland (US$73), and Norway (US$53). Most carbon taxes are levied at rates far lower than the Nordic countries, lessening their incentive effects and limiting the speed of shift of market expectations and behaviours.

Increasingly, countries are opting to impose carbon taxes as well as or rather than creating new cap-and-trade schemes. Of the recently announced carbon pricing mechanisms, nine are carbon taxes (Argentina, Chile, Colombia, Liechtenstein, Mexico, New Brunswick, Newfoundland, the Northwest Territories, and South Africa) and only one was an ETS. It appears countries and their national treasuries are realizing that the ease and simplicity of a carbon tax appeals more than complex ETSs, which can be less effective and subject to abuse and misuse.

In the next three decades, all of these (and many yet to be enacted) carbon pricing schemes will have to be increased progressively and will have to gradually converge towards a higher rate per ton if they are to move us towards our net-zero goals.

With many countries having taken the step of initiating a carbon tax or price, measured gradual increases can and should be agreed, converging at a higher level by 2030 at the latest and continuing to rise from then on through 2050. The process, if transparent, debated, and well understood, can be effective. Taxation for the common good is not an anathema. It can and should be pursued.

We can all learn from the Swedish example of how rising carbon taxes work in practice (see Box 3.1).

BOX 3.1 SWEDEN'S LEADERSHIP ON CARBON TAXES

Sweden provides the most notable example of a country applying a low-to-high carbon tax effectively over the long term. Sweden began implementing a low-to-high carbon tax in 1991. The country's carbon tax was one of the first in the world. Today, the tax is approximately US$130 per ton of CO_2. It is applied to almost all sectors of the economy. The tax was introduced as part of a major overhaul of taxation that also simplified and lowered labour taxes, placed a value-added tax on energy, and provided some state aid for certain sectors. The carbon tax was introduced at a low level and gradually increased. The result has been good for both Sweden and the planet.

With such a high tax, oil and coal use has plummeted. By 2030, emissions from domestic transport in Sweden will have fallen by 70 percent compared to 2010, and by 2045 Sweden will have no net emissions of GHGs.

The carbon tax has not been a drag on the Swedish economy, which has in fact performed strongly throughout this period, anti-tax rhetoric notwithstanding. The tax, as expected, spurred innovation, increased energy efficiency, changed business practices, and altered incentives to reward those firms that practice environmentally effective strategies while pursuing parallel business goals and to punish the laggards. Sweden's carbon tax has pushed individuals and firms to change their behaviours and practices. Today, as a result of the high real cost of carbon, Swedish firms often lead the world in application of carbon-neutral, closed-loop technologies.

Is Sweden poorer as a result of its forward-thinking carbon policies? No. Stockholm continues to be a rich capital city in a rich country. Sweden has low levels of inequality and poverty and a high degree of upward mobility.

Is Sweden less equitable because of the tax? No. Care was taken to ensure the tax did not unduly burden the less well off. Today, in 2021, Sweden is on a sustainable economic foundation and one that will help ensure future generations, not only those in Sweden, have a better chance of a liveable Earth. Not for the first time, the Swedes have shown the rest of the world how to construct a society with a stronger social contract and a more sustainable economy.

The lessons from Sweden are clear. Well-planned, transparent taxation of carbon applied over a gradual glidepath can be successful in shifting incentives and practices across a society without adversely impacting societal cohesion and economic prosperity. The exemplary Swedish case for effective application of a carbon price as a tax should be emulated. It is simple, fair, and uses market mechanisms to spur changes across industries and communities. Unfortunately, unlike Sweden, many key polluters fail to price carbon appropriately.

The Swedish case also shows how carbon pricing can operate smoothly over decades. However, while Sweden has led, most other countries have done far too little and still lag today on pricing. This delay and failure to price carbon appropriately to get the desired market incentive effects means that the price applied today, and tomorrow, must be that much higher than it would have been had more states acted on the message in Al Gore's *An Inconvenient Truth* (2006), nearly 15 years ago.

Sweden has demonstrated that countries can decarbonize without damaging their economy and while achieving GHG emission targets. It has demonstrated how to tax carbon, progressively raising taxes to a high level as businesses, investors, and individuals adjust. The Swedish economy has performed as well or better than many of its peers since the tax came into force. At the same time, the Swedish economy has evolved into a green economy without exacerbating societal inequality, as poorer families are recompensed for higher costs, avoiding social tensions. The carbon tax has succeeded in shifting incentives and redirected investment in the economy towards the needed industrial regreening and renewal. Swedish firms today are among the world leaders in operationalizing the circular economy. The know-how and products of Swedish firms are valuable and exportable. Sweden is a trailblazer on carbon taxes and offers many lessons for other countries.

Canada also illuminates a way forward on carbon taxation policy design and rollout, as explained in Box 3.2.

BOX 3.2 CANADA'S CARBON PRICE BEST PRACTICE

In 2018, Canada passed a federal carbon pricing law aimed at reducing GHG emissions. The Greenhouse Gas Pollution Pricing Act was designed to not interfere with provinces' pricing but to ensure each province's system met the requirements of the federal law. In provinces that did not act, the federal scheme would kick in instead. The law required a tax on fuels starting at CAN$20 per ton of CO_2, rising to CAN$50 by 2022. The federal government recognized this was not quite enough, so in December 2020 it announced a glidepath of progressive increases in the price of carbon through 2030. That is, the price will increase by CAN$15 each year through 2030. From then on, it will rise by CAN$15 each year afterward. The carbon price will reach CAN$170 per ton of CO_2 in 2030.

Canada will eventually be pricing carbon at or above Sweden's level and will become a global leader on carbon pricing by 2030.

In line with the recommendations of the Carbon Leadership Council, Canada rebates the entire tax to its citizens, with 90 percent going to individuals and 10 percent to businesses and institutions that cannot pass on increased costs (like hospitals). Individual payments will be scaled progressively, based on income. The government estimates that 80 percent of households should receive somewhat more than they paid in tax. Meanwhile, Canadians who have the largest GHG footprints will pay more. This progressive response is ethical, moral, and appropriate in that it addresses environmental and equity concerns, and it is the kind of policy success that can be achieved when industry lobbyists have little if any role in the process.

Researchers suggest that, in 2030, the carbon price would add about CAN$0.38 per litre to the cost of petrol (at present, the average is CAN$1.00 to CAN$1.20 per litre). The rebate payments in 2030 would range from around CAN$2,000 to CAN$4,000 per year for a family of four.

Canada has also announced that the government will explore the potential of border carbon adjustment – a type of import tax meant to protect domestic industry from goods produced in countries without similar carbon taxes. The government believes that Canadian firms should not be penalized for taking the right path while others freeload off their net-zero carbon pricing policy.

Canadian policymakers appear to be listening to Mark Carney, former Bank of England governor and current advisor to COP26 and to the technocrats: Be ambitious. Announce the glidepath. Set a gradually rising price over time. Be clear and set interim targets. Be credible. Be predictable.

Taken together, these steps will bolster market certainty and predictability and will pull forward market decisions and shift market expectations, all of which will speed up the rate of transition and lower the overall cost as more and more actors in Canada align with the goal and embed it as part of their business planning and strategy decisions.

From a climate action perspective, Canada has leapt into the policy implementation and design lead. These carbon pricing policies align with what is needed and are policymaking and implementation done properly. Others would do well to emulate this design approach.

It is worth observing that the announcement of a progressive increase in the price of carbon did not cause Canadian markets to collapse or investors to flee. Instead, they can now plan with clarity and seize the opportunities this presents as well as manage risk.

In 2020, there were 32 carbon taxes being levied – 25 by countries and 7 by regional governments. Unfortunately, the taxes cover only 3 gigatons of CO_2 equivalent ($GtCO_2e$), or 5.6 percent of global GHG emissions (World Bank, 2020).

So, to date, excluding a few successes, carbon pricing has not yet been applied widely and effectively across enough sectors and countries at high enough levels in major markets to shift behaviours in enough cases. Hence, the urgency for collective action on a minimum price (with possible carve-outs for poorer, low-emitting, countries) and a rising trajectory converging at a high level in 2030 and rising still further as 2050 approaches. A straight carbon tax is not the only option. In place of or in addition to a carbon tax, many regions and states use cap-and-trade schemes, which can be effective but also have some important drawbacks.

Cap-and-trade schemes

ETSs, which are cap-and-trade schemes, are an important part of the war on carbon. Cap-and-trade schemes operate through the establishment of regulated markets for carbon emissions permits, which ideally progressively diminish in number, which are traded by investors and emitters. Emitters must have enough permits to operate their power plants. Trades take place via regular auctions in regulated markets overseen by national authorities.

The widespread popularity of ETSs arose largely due to the notable success of the first instance of this type of regulated market – the US Sulphur Dioxide ETS, as described in Box 3.3. This early market solution demonstrated that an ETS can work and work well; indeed, the US sulfur dioxide (SO_2) market is the best example of effective market creation and operation and has been used as a blueprint by governments as they set up ETSs.

BOX 3.3 ACID RAIN, CLEARER SKIES, AND MARKETS

The success of SO_2 reductions from power plants demonstrates what can be achieved when market mechanisms are harnessed to secure ecological goals. The market for emissions was launched in 1994 in response to ecologically damaging acid rain produced by dirty oil- and coal-fired power stations that was denuding America's forests. The solution US policymakers adopted was the creation of a tradeable permit system in SO_2. Power companies were allotted emission allowances, which were subject to an annual cap, and the allowances were freely traded. Firms had to have enough allowances to cover SO_2 at each power station. The amount of SO_2 emitted by a power station was continuously monitored. At the end of each year, the power company had to have enough allowances in its account with the Environmental Protection Agency (EPA) to cover recorded SO_2 emissions from their power plants.

Importantly, the market conditions were gradually tightened to secure the desired SO_2 reductions. The scheme at first targeted the 263 dirtiest power plants, which had reduction goals set for five years. The subsequent five-year period lowered the SO_2 cap, further increasing the prices of the allowances. SO_2 allowances were freely tradeable across years. 'A relatively efficient private market developed in a few years' time', with allowances trading at between US$178 and US$205 per ton.[a]

The market operated as hoped; incentives shifted among utilities and investors, with the former closing polluting plants and the latter pricing-in future price rises. As a result, SO_2 emissions fell progressively over time and more steeply as the date of each new five-year phase was reached and market emissions allowances could be adjusted. Figure 3.6 shows power plant electric utility SO_2 emissions and market prices. Emissions fell steadily from 22 million tons in 1990 to 2 million tons in 2019, a remarkable outcome. The figure also shows the price gradually increasing and then spiking, all while we see a consistent fall in emissions, as utilities responded to the new, regulated market's signals and incentives.

Prices rose gradually and spiked during 2003 to 2006. During this period, the EPA was seeking to further cut the number of emissions permits to increase prices. Prices dramatically jumped in anticipation of this regulatory move. Prices ultimately fell back suddenly when successful utility litigation in US courts was followed by the George W. Bush administration's decision not to fight the negative court ruling with an appeal, and this resulted in gradual atrophying of the SO_2 market.

The SO_2 market is a rare real-time example of a brand-new emission market, with clear lessons for how economists and market regulators should approach climate crisis economics solutions if we are to achieve net-zero policy goals and limit global warming to levels that protect the planet's ecosystems and future generations, human and nonhuman.

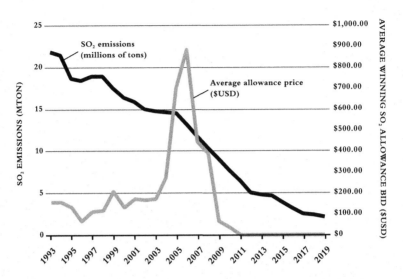

FIGURE 3.6 How the US SO$_2$ market slashed emissions and raised prices

Note: Joshkow, Schmalenese, and Bailey, 1996.

Source: EPA, 2020.

Well-designed, well-regulated markets can harness animal spirits to achieve common environmental goals. If ever there was a relatively unambiguous real-time test of the argument in favour of well-designed and well-regulated markets with clear environmental goals, the US SO$_2$ cap-and-trade market is it. This market was created in its entirety by the regulators. It was resisted by the polluters. Once up and running, it rapidly shifted incentives, prices, and environmental outcomes. Dirty power plants were closed. Emissions fell and the skies stopped raining acid on America's forests. We saw that markets, effectively designed, corralled, overseen, and incentivized, can secure environmental goals quite rapidly.

Clarity and transparency over the medium term and on the direction of glidepath are essential. Markets can deal with environmental regulatory plans. Investors prefer to know what to expect over multiple years so firms can adjust business strategies accordingly. The regulation does not need to be static; in fact, it should not be when it is seeking to achieve climate crisis goals. The US utility firms and participants in the market understood the glidepath and prices rose, and emissions fell as a result, as would be expected.

Set interim targets that are not too far off to allow for regulatory flexibility and promote business action, not inaction. The SO$_2$ market's planning horizon was not so long as to allow actors the luxury of procrastination. Actors knew the auction and yearly time frames and how they could be reappraised and emissions adjusted, and they knew there was a more significant reappraisal every five years. This approach reassured participants but also ensured market regulators could respond to changes in the market and SO_2 emissions. Utilities got certainty over the operation of the market coupled with an understanding that, over the medium term, adjustments should be expected.

New environmental markets, until established, should be sheltered from political sabotage. Had the legal and political setback seen in 2006 come seven or eight years earlier, as the market was just getting going, the outcome could have been much more negative. This can be avoided by various means, which include:

- Using existing or creating new arms-length regulatory institutions with executive power over the sector
- Legally building in a timeline that ensures certainty beyond at least one, and ideally more than one, electoral cycle
- Building support among market actors and investors to ensure resistance to political interference in the decision making on market operations after long-term goals are agreed and set.

Even significant legal setbacks are not deadly once markets have evolved and a new normal is established. The history of the SO_2 market shows that even when it was poleaxed by a legal loss and the refusal of a new, Republican administration to come to the defence of a long-standing EPA market and policy, it was not fatal to the long-term goal of lowering SO_2 emissions. Emissions continued to fall even as the market declined because by then, ten years into the market, utility companies had made the switch, changed plants, closed the worst polluters, and/or converted to gas. The market and investors had made the jump to a new normal, one that did not include burning brown coal to generate electricity. This lesson tells us that the early market formation and launch years are more important than later years when the market dynamics and shifting costs and strategies can 'lock in' climate achievements even if short-sighted politicians seek foolish and destructive rollbacks.[a]

The SO_2 market case shows that new markets can be created relatively quickly from nothing and operate effectively, driving significant environmental outcomes. The SO_2 case gives reason for hope in countries with markets that currently lack such market mechanisms.

It suggests, for instance, that a well-designed national market for GHG emissions in the US could work alongside a carbon price, if the political will to make the leap can be secured.

It also suggests that a market for carbon offsets – yet to be created – can work if designed properly and regulated effectively and can deliver GHG emission reductions, once again using market dynamics to pull forward climate change action and solutions.

Note: [a] Industry, once committed to a new architecture, will resist going backwards. For instance, the Trump administration's insistence on rolling back Corporate Average Fuel Economy emissions and miles-per-gallon standards for automobiles was opposed by major car manufacturers. The firms are already committed to them, and they are spending and planning on that basis. With California (the largest US market) refusing to roll back standards, the firms do not want to revert to earlier polluting technology. Here, we see a similar dynamic as in the SO_2 market case. This suggests that if we can lock in changes and the glidepath for a long enough period, markets will spur the dynamic and support the long-run shift.

The SO_2 market demonstrated to many policymakers and governments that ETSs could work if they were designed carefully, with allowances shrinking predictably, and with credible oversight. The US SO_2 example also illuminated how quickly change can occur when markets are harnessed, price glidepaths established, and investors and participants respond accordingly. The SO_2 case showed what is possible when politicians have the guts to agree to market solutions with aggressive goals and then leave oversight to the technocrats who know how to manage markets and ensure compliance. Unfortunately, to date, other ETSs have only gradually learnt these lessons.

As of 2020, there were ten operating regional and national ETSs covering 39 jurisdictions. These schemes cover 8 $GtCO_2e$ of global GHG emissions, or 13.9 percent of global GHG emissions. In addition to these schemes, there are 21 subnational ETSs across the globe that cover 2 $GtCO_2e$, or 4 percent of global GHG emissions (World Bank, 2020). Today, ETSs remain more important than carbon taxes overall, from a GHG perspective, even as they are levied at a much lower rate, and thus affect incentives less than is needed. Strengthening their operation must remain a key goal for GHG regulators and supervisors, wherever they are employed. Figure 3.7 highlights the ETSs in operation in 2020.

In 2020, most of the carbon pricing ETSs in major economies are not yet fully fit for purpose and must be rapidly broadened in application, strengthened, and the pricing raised and made more efficient to be reflective of the damage from GHG emissions. At present, the ETSs do not cover enough sectors or economies. The ETSs are not yet swiftly shifting incentives as needed. ETSs can be poorly regulated and are subject to abuse and misuse. The European ETS is instructive in that it is both the most effective today, in terms of levying the highest price, and yet is also a sobering example of what can go wrong.

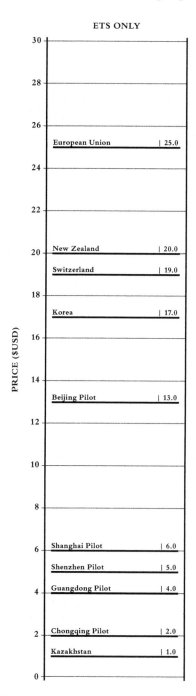

FIGURE 3.7 ETS cap and trade in operation

Source: World Bank, 2020.

Europe's ETS – the world leader, despite troubles

The European ETS is the longest-running, largest transboundary system in the world, with the largest carbon market. The scheme was created in 2005 and covers 11,000 installations across the continent, and 45 percent of Europe's GHG emissions (Carbon Market Watch, 2020). Its large market size and its scope (i.e. sectoral coverage, which includes aviation) increases its potential impact. The EU ETS enshrines the 'polluter pays' principle. However, the scheme has suffered from high price volatility and has had serious governance and monitoring failures. Past price volatility in the EU ETS has undermined market signals. For a decade after its creation, carbon emissions allowance prices did not progressively increase but instead fell gradually, precisely the reverse of what is needed (see Figure 3.8).

Why did this happen? At the beginning and in the first two five-year phases of the EU ETS operation, far too many emission permits were distributed for free by politicians pandering to power lobbies, hugely depressing ETS allowance prices. Ultimately, the European Commission had to step in and require 11 states to cut their allowances to push up prices. Prices were also adversely impacted by the ability of utilities in some states to roll over unused allowances from one year and one closed plant to another, so firms could hoard and apply unused allowances across time. Ideally, permits should lapse after a set period, requiring utilities to buy new allowances, at a higher cost.

The EU ETS also lacked effective governance and proper market supervision. Significant fraud resulted, damaging the market and undermining trust (Frunza *et al.*, 2011). At one low point, permit exchanges in that market were probably spurred more by the illegal activity that took place in the absence of proper regulation

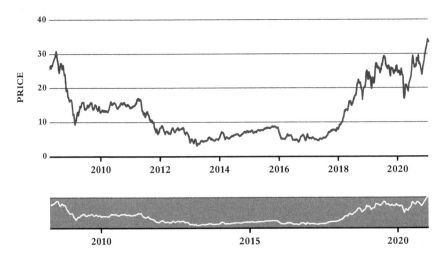

FIGURE 3.8 Is the EU seeing a carbon pricing breakpoint?

Source: highcharts.com.

than by the actual need to cover emissions (because too many permits existed) (Borghesi and Montini, 2014). Large blocks of emissions allowances also mysteriously disappeared. Issues of fraud and abuse were so rampant that the European Commission assumed direct oversight in 2013.

EU ETS market prices, supervision, and oversight failed in the first period of the scheme's operation. The European Commission had to take steps in 2019 to soak up unused allowances and stabilized prices with the introduction of an EU Market Stability Reserve. Both steps further improved matters and set the stage for rising prices and market sentiment, which were visible in 2020 (see Figure 3.8). Now the market and investors reacted to a tightening of supply and supervision, as would be expected, by raising allowance prices, which have rebounded because of this more effective regulation, supervision, and transparency.

Importantly, in 2021, carbon prices are being pushed higher still with the coalescence of a new market narrative among participants that assumes future carbon price rises driven by shifting expectations and the COP26 process. Prices in the EU ETS are now near all-time highs, a welcome development, although they are still too low to achieve net-zero goals. If the EU ETS has a sorry early history, now improved by better oversight, rising prices, and the prospect that a market narrative shift is supporting price rises, China's ETS has a long way to go to deliver European-scale results.

China's ETS – a lot of room to grow

China's ETS finally began in 2020 and is not yet fully effective, but there are some small positive signs, from their pilot programmes, that may be turbocharged by President Xi's announcement of the country's net-zero goal. The China ETS put a price on carbon emissions, creating modest incentives for companies to cut pollution. China announced the national scheme in 2018 and spent two years building the reporting system and market infrastructure. In the first instance, it covers coal-fired power stations – among the most egregious polluters. The scheme's early operation and oversight will be a crucial test, as the IEA (2020) notes:

> The initial years of operation will be crucial to test the ETS's design and establish trust. Given the dominance of coal power in China's power sector and in its overall CO_2 emissions, how the country's fleet of coal-fired power plants is managed will be essential for China to meet its climate goals and other sustainable energy goals.

The Chinese are not placing all their carbonized eggs in one basket. The authorities also aim to create a market in carbon derivatives to further support the ETS. Eventually, the China ETS should cover one-seventh of global GHG per year and 100,000 industrial plants once fully operational (IEA, 2020). Chinese authorities are learning from others' failures and are introducing a carbon tax and renewable power quotas to also speed the shift in incentives. They understand it is not a matter of one lever or one policy but both and more. China is lagging in 2021, but other mechanisms are being put in place and environmental goals set across all sectors.

The China ETS will test whether Chinese state capitalism can respond faster than Western-operated schemes. It may be the case that direction and supervision from the powerful centre can have immediate, direct effect on business actions and outcomes across many regions and locations. China does not have to respond to public complaints over the price of carbon; for instance, we will see no *gilets jaunes* protests in China. The next five or so years will be the real test. Will the Chinese ETS regulators control emission permits and cut the supply? Will they lay out a clear, predictable, credible plan and execution, shifting market expectations and incentives? The SO_2 example shows how to do it. The EU ETS case shows the mistakes that can be made. Both examples show that close regulatory oversight, shrinking emissions allowances, and effective supervision are essential to ensure markets are harnessed to operate in support of societal goals. As China is pressing forward from a low point on ETS, so too is the US, with a weak patchwork of regional, not national, schemes. This needs to change under the Biden administration.

Whither the US on carbon pricing

The Biden administration clearly wants to retake a leading role and help drive the UN COP process. The US has re-signed the Paris Agreement and must now begin to implement net-zero policies. President Biden has signalled he will restart the green revolution and reseed a GND for the US. Announcing a plan to price carbon and beginning the process of standing up the mechanisms and enforcement systems are essential policy signals that America is serious about its own war on carbon. Even in 2021, the largest economy in the world fails to nationally price carbon. Past national moves in the 1990s and in 2000 to price carbon failed in the US Congress. Getting pricing through Congress will be a huge challenge for the Biden administration and pits the new economy against the old, entrenched fossil fuel lobbies against voters and environmentalists, young voters against retirees. The fight will be vicious and not easily won. In 2021, regional programmes provide a thin and incomplete patchwork of partial, varying pricing schemes in some states. This is not enough; federal action and direction are sorely needed in addition to this state-level action.

A thin regional patchwork, with some successes

The Regional Greenhouse Gas Initiative (RGGI), created in 2009 and the oldest such agreement, applies to the utility sector and is binding among signatory states. It has an annual cap, progressive annual reductions in the cap, and an auctioning of permits to raise money for energy efficiency programmes by the participating states. The cap was reduced by 30 percent from 2009 to 2019 (climatechange.org, 2019); GHG emission by power plants in the RGGI have fallen by 47 percent in the same period and generated US$6 billion in environmental and multiplier benefits (Acadia, 2019).

The Western Climate Initiative (WCI) is a cap-and-trade system that was created in 2007. Participating states agreed to GHG targets of 15 percent below 2005 levels by 2020. Participating states are also required to adopt California's auto emissions policy. Unfortunately, anti-tax activists succeeded in forcing many states to back out of the WCI. Ultimately, California and Quebec implemented the system, which applies across most sectors, including transportation. The WCI has seen progressive downward adjustments of caps, regular auctions, and rebates to consumers, and generates US$12 billion in revenue for the sates. The cost to consumers in 2016 was a very manageable US$75 per motorist (ClimateXchange, 2018). This shrunken WCI has largely worked as planned. California's ETS partici-pation helped the state reach its 2020 emission targets by 2016 because of a rapid reduction in power plant emissions. This was driven by a rise in hydropower and falling prices for solar and wind generation, working in tandem with the cap-and-trade system.

These limited regional schemes are better than no action and still show that regulated ETSs do begin to shift incentives and can help reduce GHG emissions. The WCI has had a significant impact on emissions. Both the RGGI and WCI demonstrate positive dynamics. They have affected market incentives and outcomes within their regions to varying degrees, and they show ETSs can work. The regional ETSs have begun to make markets, firms, and individuals account for and intern-alize the real carbon cost of doing business. Crucially, they can be managed and operate smoothly. Of course, a great deal more needs to be done to price carbon than these limited schemes.

A national US ETS is needed

Ultimately, the Biden administration should establish a national, US-wide ETS, as well as a tax on carbon. This will require a nationally regulated carbon market, auctions, and relatively swift annual reductions in emissions caps, plant to plant, market to market, and sector to sector. A revenue-neutral approach should be considered, with income adjustments, to minimize opposition. Once established, with long-term goals set by Congress, a new ETS must be insulated from polit-ical interference to allow the market to rapidly grow, become more efficient, and harness market dynamics to the goals of net zero, just as the successful SO_2 market did. As we saw in the SO_2 case, markets can take root if there is credibility, predict-ability, transparency, and effective supervision.

Learn lessons and apply them globally and locally

Governments should learn the lessons from Sweden, Canada, the US SO_2 case, and the operation of the EU ETS (both good and bad) and apply both carbon taxation and cap-and-trade levers to the task of achieving net zero. Many governments already apply both mechanisms. More governments should apply one or both to help speed and secure their announced net-zero goals.

Thus far I have avoided stating what the baseline minimum carbon price should be. It is time to come off the fence and clarify what we should be paying to internalize carbon polluting and shift markets, practices, and behaviours.

Carbon prices: How high is high enough?

Just how much higher must global and national carbon prices be per ton to speed change and help achieve net zero by 2050?

- Higher than Nordhaus's latest DICE so-called 'optimum' suggests, which is US$37 a ton. This is not optimum – it does not achieve GHG and Paris Agreement temperature goals and should be rejected as too low.
- Higher than the Carbon Leadership Council suggested, at US$40 a ton. This price may have been ambitious four years ago, but America has done nothing in the interim. We need to be more ambitious.
- We need a price rise approaching and exceeding Sweden's price of 114 euros per ton and upward. We know from Sweden, and now Canada (and a few other cases), that such pricing is not the end of the economy but rather a reseeding, retooling, and reimagining.
- We should set a global price minimum of US$40 by 2022, rising to over US$100 by 2030, and continuing to rise gradually to US$300 by 2050, in an orderly shift to net zero, as recommended by the central bankers of the world.

The Network for Greening the Financial System (NGFS) lays out two possible carbon pricing scenarios: an orderly transition and a disorderly transition in which net-zero and temperature goals are met or not met. Each scenario depends on the strength of the response by governments, central banks, economies, and societies, as shown in Figure 3.9. Transition and physical risks increase as we get further away from orderly to disorderly, and from meeting to not meeting collective climate change goals.

The NGFS estimates that pricing carbon to restrict global temperature rises to well below 2 degrees Celsius in both the orderly and disorderly scenarios is the more sensible option. The NGFS members are not revolutionaries; they are conservative. Yet the NGFS's analysis in 2020 led them to conclude that the carbon price will have to be high and rising, as per Figure 3.10. Governments at COP26 and beyond should adopt the NGFS approach and be ambitious, set a minimum, and raise the price, year after year, to decarbonize the world.

The NGFS orderly scenario raises carbon prices from 2020 in a gradual manner to US$137 by 2030 and US$300 by 2050. A plurality of the world's central bankers, including from China, France, the UK, and scores of other countries, have estimated this is the carbon price needed to achieve our goals. Governments need to act.

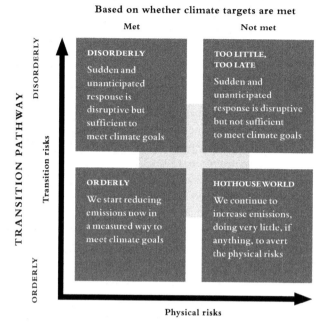

FIGURE 3.9 NGFS transition scenario matrix

Source: NGFS, 2020.

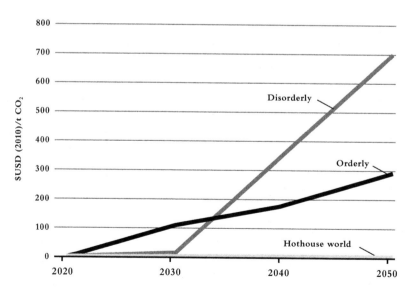

FIGURE 3.10 NGFS carbon pricing trajectories

Source: NGFS, 2020.

In both scenarios, central bankers reject Nordhaus's DICE low carbon price estimates because they would result in terrible global climate outcomes. Appropriately, central bankers are estimating what carbon price trajectory is necessary to achieve the stated policy goals of limiting global temperature increases to 1.5 degrees or well below 2 degrees Celsius.

The NGFS position is a credible and essential price path that national authorities should adopt at COP26. If governments fail to act, the cost of failure will be high. In the NGFS's disorderly scenario, nothing is achieved at COP26. The central bankers assume a sudden breakpoint comes in 2030, at which juncture a disorderly rush to respond and price carbon prompts a much more abrupt step change and steep rising carbon pricing from zero to US$700 per ton in 2050.

The message from central bankers on confronting the climate crisis head-on is clear: Act now or pay a heavy price. Central bankers are urging leaders to price carbon much more in line with what Stern maintained was necessary in 2007. But as we saw in Chapter 2, Stern's position, and that of Wagner and Weitzman (2015), was drowned out by the ethically dubious high-discount rate and little-or-no-action stance of those more interested in neoclassical models and assumptions than human and nonhuman species survival.

Just as the NGFS central banking community has made a leap in its thinking on carbon price, so too must policymakers, businesses, economists, and individuals. Leaders convening in 2021 must decide: Pay a price today or pay much more later. Pricing carbon, by whichever means, is not avoidable.

2021 will be our leaders' breakpoint, a possible narrative tipping point. Can they deliver?

COP26 and carbon pricing's tipping point

A breakpoint, a narrative tipping point is building for the net-zero transition and carbon pricing, for the EU cap-and-trade system, for US reengagement and leadership, and for collective action, with burden sharing and enforcement. In 2020 and 2021 – a positive market and narrative tipping point – is underway. Covid-19 in early 2020 caused the EU ETS prices to fall, but they have bounced back to near all-time highs, despite the pandemic raging on through 2020 and into 2021. Prices rose despite lower energy demand and a collapse in economic activity. Why might this price rise be happening? European ETS markets appear to be factoring in carbon price increases they expect to see from COP26 and are planning accordingly. They are telling themselves a new story about the future, one that integrates climate change policy shifts into their thinking and actions.

A new narrative is developing across many businesses and sectors, one that posits that progressive increases in carbon and environmental standards should emerge from COP26. Businesses are anticipating a pricing of carbon and a tightening of carbon markets and rising costs. They are reassessing strategies and embedding net zero in their corporate goals in advance, looking ahead to 2050. Firms are pricing carbon higher because they expect governments and markets to increasingly reward first movers and further punish polluters. Hence, premiums (for polluters wanting loans) and leaders are making the shift (in terms of their equity valuations).

It is imperative that government leaders seize this opportunity and speed market shifts to a permanent decarbonization narrative. If governments do so, the transformative dynamics of market momentum can be swiftly increased and strengthened. Shiller (2019) shows us that in economies and markets, narratives drive decisions, mould investors' views, and can be built upon or destroyed by events and setbacks. Leaders at COP26 can support the climate crisis narrative shift and deliver pricing signals that confirm these already evolving market sentiments. Seizing the chance to send the right signals can underscore decisions such as BP's move to write off tens of billions of UK pounds in oil reserves it anticipates will become stranded assets. COP26 action would underline and reaffirm this shifting narrative. Doing so would punish others, such as Exxon, which talk the talk but fail to walk the walk. Government action needs to underscore the decision by leading banks and sovereign wealth funds, such as France's Société Générale or Singapore's Temasek, that increasingly disinvesting in coal or moving out of oil and fracking can be indirectly supported. Acting on carbon pricing would send the signal that pension fund disinvestment from coal and fossil fuels – such as that announced by the New York Pension Scheme – is appropriate and ought to be replicated by others. Government action and clarity on the net-zero pathways buttresses those leading the change. It implicitly supports decisions such as those by BlackRock to invest heavily in the UK EV market and technologies. If more governments follow up stretch goals on GHG emissions with announcements on EV infrastructure (such as those expected from the US), this again confirms the narrative shift, alters market decisions, triggers investment shifts, and changes sentiments.

Central banks can add to the narrative and commercial pressure by changing macro and micro supervision of climate, transition, and physical risks. Central banks should also, through their technocratic forums, signal the eventual change in capital weighting to reflect the increased risk in lending to carbon-intense industries.

In recognizing the narrative shift taking place via action and announcements, governments and their regulators will affect the market positions and the equity values of increasing numbers of investors who are committed to renewables, to new markets in offsets, and to innovations that are greening the economy. This climate transition story can be further strengthened and reconfirmed. As with the climate, these signals and steps are interconnected in a linear and nonlinear fashion. Government actions and clarity on carbon pricing mechanisms and implementation at COP26 and beyond can and should drive the policy and public narrative in many parts of the world to a series of tipping points that support economic sustainability, resilience, redesign, and reimagining and the start of a new, green industrial revolution. We are rushing towards our own story telling and policy tipping point. It is policy versus planet – action versus inaction and disaster. If we get there in time, we can arrest global temperature increases and secure our survival and that of other species.

Momentum always matters. A virtuous cycle of network effects and feedback loops of the type the planet needs is possible if governments seize the chance to reconfirm and underscore the urgency of the climate crisis and the direction, glidepath, steepness, and plans for the carbon price narrative shift that has begun. The speed of transition to the net-zero future can be accelerated. As Tooze (2020b) notes:

> Imagine if speculators persuaded themselves that the smart thing was to be in on the realization of [climate change] policy, thus hastening that outcome and making it easier for governments to stay the course. Remarkably, that is what appears to be happening in EU carbon markets.

Markets are looking at the climate crisis and trying the read the trends and direction. It is essential therefore that COP26, the EU, China, the US, and others continue to send the confirmatory signals. If this occurs, and it just might, especially with a new US administration that believes in science, facts, and climate data, this long-awaited positive planetary climate change narrative shift and tipping point can be built upon. Many businesses are engaged on the net-zero journey already and they are looking for the right regulatory signals from the world's governments. COP26 negotiators in Glasgow must give it to them.

A great many CEOs I speak to are serious about the net-zero transition and their role in making it a reality. These leaders, after all, have children and grandchildren. They want to leave a meaningful legacy, not one of destruction and debasement of the planet. CEOs and their employees want to be proud of the businesses they work in and the social and economic role they play in this wider societal shift. This is no longer just greenwashing. Sure, some firm's CEOs are just talking the talk, but an increasing number are walking the walk and are engaged in a real change in mindset and business strategies, which can help speed the desired outcome.

Businesses and leaders that are making a leap as first movers and risk takers still require the right confirmatory policy signals. The firms on the leading edge taking short- and medium-term economic risks to achieve net zero rapidly, innovating and redesigning, should be supported through policy clarity, credibility, and predictability. Markets will then increasingly reward firms that are delivering and innovating and punish those that are in denial and polluting the planet. We need to see a series of positive feedback loops supporting the climate change policy goals in 2021 and beyond. This is not assured, and there is plenty of scope for politicians to fail us as they have done again and again on climate change. Failure at COP26 would be a massive setback and markedly increase the likelihood of much more abrupt changes in carbon prices a few years hence, when the climate and terrible unfolding scientific reality can no longer be avoided. So, what should progressive leaders do if no acceptable, suitably ambitious consensus is reached in Glasgow?

Can we get agreement at COP26?

It would be naïve to assume a global consensus on a meaningful and rising carbon price will be easily achievable at COP26. It is quite possible there will be no global agreement if Brazil, India, Russia, and others balk. Getting to yes requires leadership, strength, common purpose, and a clear sense of national and international collective responsibility. Farsighted leadership is needed but may well be lacking in those countries that are unable or unwilling to pull together for the common good.

American populism still stalks the electoral landscape, even if Trump lost by a landslide in November 2020. Can the Biden administration swiftly take up global leadership once again and pass a GND at home plus a deal on climate change and a carbon price globally? Or might he be bullied into weaker, insufficient action, harried by lobbyists, a split Senate, and domestic sectors blinkered by their short-termism and selfishness? It is too soon to judge. The president's plan, announced in 2020, calls for spending at least US$2 trillion, with the goal of moving entirely to renewables by 2035 and achieving net-zero emissions before 2050. If the plan is implemented, this will help the US speed the transition. If he can get the plan through Congress, it will extend and reinforce market dynamics already being seen in Europe and elsewhere. America is coming back into the diplomatic fold, returning to cooperative leadership, rejecting Trumpian climate denial. All these signals shift expectations and alter business planning and strategies. This is hugely significant for the globe. Yet even if President Biden takes a renewed and much-needed leadership role in the runup to COP26, what of the positions of other major states?

Britain is led by Prime Minister Boris Johnson, who is chronically unable to make hard decisions, is prone to lying, and is poorly prepared. Is Johnson in a position, politically or strategically, to make Britain's net-zero leap even more ambitious with changes to the net-zero plan and a tougher timeline and carbon pricing commitment alongside it? Can Britain's Brexit prime minister ensure the future of England's green and pleasant lands and sign an ambitious deal in Glasgow? It still seems possible. Johnson is being pressed by Mark Carney, as advisor to the UK COP26, and as UN Special Envoy on Climate Action and Finance. Perhaps Carney can create an opening for Churchillian climate change policy greatness for Johnson. Let us hope he seizes it.

India's current populist prime minister and strongman, Narendra Modi, should seize the chance of a leadership role as India faces some of greatest emerging risks of all nations from climate change. Its people are extremely vulnerable to adverse weather effects. Yet if India's past negotiation history is any guide, it is likely Modi will resist a climate change deal he so dearly needs because of nationalist notions at odds with the global scale of this crisis and its necessarily collaborative solutions.

Russian President Vladimir Putin may also resist a meaningful climate change deal and carbon price, which his extractive, poor, backward, commodity-driven economy would find hard to bear. Could the Russian Bear block success? It is indeed likely. No one calls Putin a nascent environmentalist; he has done little if anything as Siberia has erupted in flames and as methane substrates have bubbled to the surface of the Russian Arctic Ocean.

What of Turkey and Brazil, which are also led by populist nationalists? Will their leaders recognize the need for sacrifice on climate change and carbon prices for our common good? It seems highly unlikely, with the former sabre rattling in North Africa and the latter allowing his country's most precious asset, the Amazon forest, to burn and be denuded and desiccated.

What of Africa and lower-income countries, which rightly demand much larger transfers than the promised (but not delivered) US$100 billion a year to address issues of justice and equity and the transition? Will creditor advanced nations deliver at a time when they are stressed by Covid-19, when they have yet to live up to past pledges of annual support? And absent meaningful resource transfers, can we expect the very poorest to join the carbon price push? Can they afford to do so? Perhaps not.

These are only a few illustrative major geopolitical and national political stumbling blocks that confront government negotiators as they seek to flesh out net-zero and carbon price plans as elements of a climate change deal that works for the many but that has costs and possible future economic benefits for all.

I believe that rather than be discouraged, we need to recall that the history of major shifts and leaps is one of slow gradual steps and changes, until a crisis is really understood and a new consensus and narrative tipping point is reached. Then, quite suddenly, a move from one equilibrium to another occurs, just as we also observe in other crises and in the natural world, of warming and of ice ages and sudden extinctions. As Nelson Mandela observed in a 2001 speech, something 'always seems impossible until it's done'.

The political, policy, and economic inertia before a tipping point and narrative shift can seem immense, until it is suddenly overcome. It is still possible the climate crisis can be recognized and that a meaningful shift can occur with farsighted leadership and pressure. Compromise now and a longer glidepath is, after all, in everyone's planetary best interest. It is still possible this crucial juncture might be reached in 2021 and we will see a political and diplomatic breakthrough at COP26 and then begin to see change implemented and enforced.

Or perhaps not. If no deal is possible, or if there are laggards who drag their feet and fail to live up to their commitments, if a disappointing and weak outcome looms large, those leaders committed to action must instead construct a coalition of those prepared to do what is necessary to protect the climate as a global public good, to address the tragedy of the commons that climate change represents, and stop freeloaders and polluters from prospering now only for future generations to pay the price.

Construct a coalition of the willing

Given the difficulty of reaching a global consensus on how to proceed with enforcing the next steps to net zero via carbon pricing and scores of other interlinked incentives, penalties, and mechanisms designed to further bend the GHG curve, willing and more committed COP26 leaders must prepare to proceed as a coalition of the willing, excluding those who refuse to act.

Those that agree on pricing, glidepath, implementation, and monitoring can and must move ahead and begin to enact necessary changes to policy and incentives; they do not have to wait for everyone to agree. This approach, which Nordhaus calls the 'club approach', permits speedier progress precisely because participants in the group exclude the free riders and do not require complete global unanimity. The exclusion allows fast action absent global consensus. In 2021, that may be where climate change activists in governments and outside find themselves.

This approach was used in the creation, growth, and strengthening of the EU over 70-plus years. Only those who agreed to adhere to democratic and federalist norms needed to apply or would be admitted after negotiations. Once inside, participants got full membership benefits including access, pricing, common regulation, and oversight. The EU has been a remarkable success in large part because it is not universal and is not entirely consensual. In the EU, this enforceable, coordinated progress left naysayers and underperformers behind.

This coalition approach allows others to go their own, backward, self-destructive way – just as the UK has done in its exit from the EU. A coalition of the willing does not require others to wait or halt their progress. This can be seen in the NGFS example (see Box 3.4).

Crucially, those inside a coalition do not have to treat those outside in the same manner. To the contrary, those outside the coalition should pay a significant cost for being freeloaders beyond the boundary of the new carbon pricing group's jurisdictions; some more than others, some less, depending on their level of development, for instance. Such a coalition can expand or contract, but generally effective dynamic associations of nations expand as those outside increasingly clamour to get in and become part of the process and decision making.

BOX 3.4 THE NGFS CASE – AS A COALITION OF THE WILLING

The coalition approach is already being successfully used in the climate change space. The 2017 creation of the Network for Greening the Financial System (NGFS) shows this. Mark Carney, then Bank of England governor, joined by allies in the Banque de France, the Dutch Central Bank, the Peoples' Bank of China, and others, created the NGFS to lead the shift on climate change and in the role of central banks and supervisors. In only three years, it has grown from 8 founding members to 83 central banks and regulators members and 13 observers.

The NGFS leads on climate change analysis for central banks, spurring action and leading and feeding into other forums and processes. The key here is that they did not wait for a recalcitrant America gripped by isolationist anti-science furore. The NGFS members did not wait for a timid US Federal Reserve Bank, which only in December 2020 joined the NGFS but still has taken no public steps or issued regulatory guidance on how to address climate change risks in America's financial system and largest banks. The US Federal Reserve is not alone in its intransigence. The US Securities and Exchange Commission is likewise somnolent and unwilling to force companies to begin disclosing climate change plans and glidepaths to net zero. Canada's banking regulators, all too close to banks that finance tar sands investments, have also taken no steps.

The coalition approach has allowed 71 central banks and regulators to begin to address climate change risks much faster and disregard blocking players and denialists to address the climate crisis ahead, galvanized by the early leadership from Carney and ongoing commitments to change from scores of major institutions.

Construction of a coalition of the willing does not negate the necessity to push for a deal at COP26 and beyond, to try to craft a universal deal and consensus, but we know consensus is difficult and sometimes impossible to secure, and we also know that consensus can lead to the lowest common denominator and a deal that does not deliver on net-zero goals.

That cannot be an acceptable outcome, and leaders need a plan B, an alternative scenario that is realistic and that can pressure laggards to commit to real action. That plan B should be the coalition approach and a willingness to move forward aggressively in 2021.

If an ambitious deal is not possible in Glasgow, the 'time has come for Europe, the US and possibly China to create a global "climate Club"' (Wolff, 2020). This coalition of the willing can move forward on carbon pricing and more aggressive climate change emissions reduction monitoring and enforcement. The leaders who still seek to push Gaia to the brink and beyond should pay a price and be forced to bear a meaningful and sustained economic burden until they come to their environmental senses.

Free riders should be punished with a carbon border adjustment tariff

Noncompliant countries – those that do not implement and enforce a rising carbon price or tax or that do not present proposals to implement net-zero plans – should pay a carbon border adjustment tariff to export and compete with businesses in countries where all are indeed paying the real price of carbon and working to be better stewards of the environment for future generations and achieve net zero by 2050.

The coalition of the willing should apply a carbon border tax against laggards and polluting nations' exports. European Commission President Ursula von der

Leyen has repeatedly argued for a carbon border adjustment on carbon-intensive imports to prevent production being shifted abroad. Carbon border adjustment can be implemented in line with World Trade Organization (WTO) rules.

Talk of a border tax adjustment tariff is anathema to neoliberal free traders and proponents of globalization the world over. They warn this would slow trade, undermine growth, and punish people as well as governments. Perhaps. Nonetheless, we cannot strive to rapidly address net zero if we continue to permit free riders and their abuse of the global public good and global commons. A coalition of the willing cannot permit states to drag their feet and fail to adhere to climate change goals and scientific evidence at no cost with no penalty. That is not free trade; it is imbalanced, unfair trade. As Wolff (2020) states:

> A club including the three major economies of the world would make it difficult for any country to free ride on climate mitigation … Conditions have never been better to negotiate an effective climate club.

As we have seen, Europe, the US, and China have already agreed new net-zero goals. Let us assume they move on with implementation, with a minimum carbon price and increases, and enforcement. It is simply not 'free and fair trade' after that to permit companies exporting from Russia or another denialist state the same low duty or practically duty-free access to European, US, and Chinese markets and so to compete unfairly against firms applying a realistic carbon price to their products and processes. When coalition members react by requiring a border tax adjustment for goods coming from states and from firms that fail to comply, this is not a denial of globalization. It is Green Globalization 2.0.

Could it be done?

Application of a green border tariff would not be especially technically difficult. Customs authorities already apply the harmonised system of tariffs to enforce WTO norms and rules. The coalition could lift this approach and apply it to speeding the net-zero transition while protecting states and firms that are making the leap – at some cost – and punish those states and firms that choose not to. Rather than damaging globalization and trade, a border tariff focuses policymakers' and businesses' minds on the trade-offs they make in acting or refusing to act, a trade-off we have hitherto avoided.

Applying a green border tariff against denialists and laggards would alter the flow of trade, shifting the composition of the winners and losers away from polluting states and providers and towards others that comply with the new rules and norms. That is precisely the effect we would want to see – changes in political, business, and market incentives decisions and behaviours. A green border tariff would be an effective statewide economic nudge to denialists to change policies and get with the net-zero programme.

States in the coalition could use the approach used in the WTO, membership in which is sought precisely because being outside penalizes countries. A green border tariff applied with a most-favoured nation approach to all in the green coalition of the willing would benefit insiders and punish outsiders.

A green border tariff could also usefully include special carve-outs for lower-income, lower-polluting countries pursuing the shift from a low GHG level and so support a just transition rather than undermine that goal. What we need is a trade tax mechanism alongside carbon pricing to punish states that refuse to agree and to enforce carbon pricing and a rapid reduction in GHG emissions en route to net zero by 2050. As denialist countries shift and begin applying recognized metrics and appropriately rising carbon prices to their markets and producers, they could apply to join the coalition and agree to the common rules and strictures, just as WTO states and the EU do with applicant member states today. The pressure to confirm and achieve net-zero goals would build, accelerate, and harness markets and reward those who were taking measurable, consistent, enforceable steps towards agreed common goals.

Enforcement will be necessary for consensus or coalition

Even if COP26 and its process achieves a radical green consensus, enforcement and penalties will be necessary. If a coalition approach is used, enforcement will also be required. After all, agreements without review, assessment, and action against the delinquent often fail to be respected. This is precisely why the WTO has just such an enforcement architecture and why the EU has the European Court of Justice to interpret and enforce the rules across the continent. Arguably these two courts have been the most effective guarantor, enforcement mechanisms, and drivers of regulatory rigour, ensuring greater and fairer globalization on the one hand, and political economy and regulatory strength on the other. Governments need to construct similar enforcement and review mechanisms to support pricing and carbon market operation. They should ultimately delegate the task of reporting and review to the technocrats below them, who are better able to oversee and supervise markets and firms (Chapter 5 addresses the institutional constructs that can help achieve this).

Strike the deal, delegate national market regulators to act, and report and review

If COP26 can set the stage for carbon pricing, it should leave to others or delegate to others the responsibility to apply the taxes, set the pricing reforms, and report upwards to the UN process and via existing mechanisms. These crucial mechanisms need greater credibility and effectiveness than the UN can provide alone. COP26 leaders need to recognize that the UN climate change negotiations process, while delivering consistently (if slowly) as a diplomatic and political process of negotiations, is poorly placed to handle carbon market regulation tasks.

Shifting incentives across all sectors

Bending the curve on climate change requires adjusted carbon price incentives across all industrial sectors and activities. This has to a limited degree begun. But it is still too slow and limited to achieve our common global goals. Yet shifting incentives across all sectors is essential. One sector cannot free ride off another. If utilities and users must pay the cost of carbon, so too must drivers, airlines, airline travellers, farmers, consumers, and shipping companies.

This point appears so obvious as to be anodyne. Yet in 2021, climate change and carbon pricing policies are still a threadbare patchwork with too many holes. Too many of us avoid carrying a fair burden and paying the real planetary price of pollution. It bears restating. Bending the climate change curve requires us to change incentives across the board and pay increased costs, and to seize new opportunities to secure a liveable Earth for future generations. Prices will rise in many areas. This may be challenging but can be addressed to ensure equitable outcomes that protect the poorest and place the largest burden on those best able to bear it. Pricing can change, but so too must markets, preferences, and individual behaviours.

Note

1 A circular economy is one in which firms and organizations seek to avoid as much pollution and waste as possible by designing out waste and pollution from processes, and in so doing lower emissions and adverse effects on the climate and the environment. Swedish firms are leaders in this circular design process; the country's tax on carbon drives firms to strip out polluting and wasteful systems – to close the circle – thus becoming more efficient and ethical, and more environmentally mindful of the impact of decisions.

References

Acadia. (2019) 'The regional greenhouse gas initiative: 10 years in review' [Online]. Available at: https://acadiacenter.org/wp-content/uploads/2019/09/Acadia-Center_RGGI_10-Years-in-Review_2019-09-17.pdf (accessed: 22 September 2020).

American Action Forum. (2021). 'The American jobs plan and the Green New Deal' [Online]. Available at: www.americanactionforum.org/press-release/new-podcast-episode-the-american-jobs-plan-and-the-green-new-deal (accessed: 17 May 2021).

BBC. (2019) 'EU carbon neutrality: Leaders agree 2050 target without Poland', 13 December [Online]. Available at: www.bbc.com/news/world-europe-50778001 (accessed: 19 October 2020).

———. (2020) 'Prince Charles calls for "Marshall-like plan" to combat the climate crisis', 21 September [Online]. Available at: www.theguardian.com/uk-news/2020/sep/21/prince-charles-climate-crisis-marshall-plan (accessed: 22 September 2020).

Borghesi, S. and Montini, M. (2014) 'The European Emission Trading System: flashing lights, dark shadows and future prospects for global ETS cooperation'. In *The EU, the US and Global Climate Governance*, edited by C. Bakker and F. Francioni. Abingdon, UK: Ashgate Ltd, pp. 115–125.

Breeden, S. (2021) 'Bank of England tells banks to brace for sky-high carbon price', 14 January. *Bloomberg* [Online]. Available at: www.bloomberg.com/news/articles/2021-01-14/bank-of-england-says-prepare-for-carbon-prices-to-triple-to-100 (accessed: 19 January 2021).

Carbon Market Watch. (2020) 'EU emission trading system as an important tool to achieve the objectives of the green new deal', 23 June [Online]. Available at: https://carbonmarketwatch.org/2020/06/23/the-eu-emission-trading-system-carbon-pricing-as-an-important-tool-to-achieve-the-objectives-of-the-green-deal (accessed: 29 January 2021).

ClimateChange.org. (2019) 'Climate change ambition renewed in 2019 after a decade of decline', 12 December [Online]. Available at: https://climate-xchange.org/2019/12/12/cap-and-trade-ambition-renewed-in-2019-after-a-decade-of-decline (accessed: 22 September 2020).

ClimateXChange. (2018) 'Regional cap and trade: Lessons from the Regional Greenhouse Gas Initiative and Western Climate Exchange'. ClimateXChange, Boston, October 2018 [Online]. Available at: https://1akqm23qb5w51pwn3n2deo7u-wpengine.netdna-ssl.com/wp-content/uploads/2018/08/Cap-and-Trade-Report-10.03.2018-compressed.pdf (accessed: 23 September 2020).

EPA. (2020) 'Acid rain program' [Online]. Available at: www.epa.gov/acidrain/acid-rain-program#overview (accessed: 17 May 2021).

EU Commission. (2018) 'A Clean Planet for all: A European strategic long-term vision for a prosperous, modern, competitive and climate neutral economy', 28 November [Online]. Available at: https://eur-lex.europa.eu/legal-content/EN/TXT/PDF/?uri=CELEX:52018DC0773&from=EN (accessed: 29 January 2020).

———. (2020) 'Submission by Croatia and the European Commission on behalf of the European Union and its member states', 6 March [Online]. Available at: https://unfccc.int/sites/default/files/resource/HR-03-06-2020%20EU%20Submission%20on%20Long%20term%20strategy.pdf (accessed: 29 January 2021).

Financial Times. (2018) 'Janet Yellen calls for US carbon tax' [Online]. Available at: www.ft.com/content/a70edd88-b486-11e8-bbc3-ccd7de085ffe (accessed: 29 January 2021).

Frunza, M.-C., Guegan, D., Thiebaut, F., and Lassoudiere, A. (2011) 'Missing trader fraud on the emissions market'. *Journal of Financial Crime*, 18: 183–194. doi:10.1108/13590791111127750.

Gore, A. (2006) *An Inconvenient Truth*. Emmaus, Pennsylvania: Rodale Press.

IEA (International Energy Agency). (2020) 'China's net-zero ambitions: The next Five-Year Plan will be critical for an accelerated energy transition', 29 October [Online]. Available at: www.iea.org/commentaries/china-s-net-zero-ambitions-the-next-five-year-plan-will-be-critical-for-an-accelerated-energy-transition (accessed: 20 December 2020).

IPCC (Intergovernmental Panel on Climate Change). (2019) 'Special report: Climate change and land' [Online]. Available at: www.ipcc.ch/srccl/chapter/summary-for-policymakers (accessed: 17 October 2020).

Joshkow, P., Schmalenese, R., and Bailey, E.L. (1996) 'Auction design and the market for Sulphur dioxide emissions'. NBER Working Paper 5745 [Online]. Available at: www.nber.org/system/files/working_papers/w5745/w5745.pdf (accessed: 29 January 2021).

Mandela, N. (2001) Speech [Online]. Available at: www.lionworldtravel.com/news/8-nelson-mandela-quotes (accessed: 8 February 2021).

NGFS (Network for Greening the Financial System). (2020) 'Climate scenarios for central banks and supervisors' [Online]. Available at: www.ngfs.net/sites/default/files/medias/documents/820184_ngfs_scenarios_final_version_v6.pdf (accessed: 17 May 2020).

Shiller, R. (2019). *Narrative economics: How Stories Go Viral and Drive Major Economic Events.* Princeton: Princeton University Press.

Stiglitz, Joseph E. (2019) 'Addressing Climate Change through Price and Non-Price Interventions'. NBER Working Paper No. 25939, National Bureau of Economic Research, Cambridge, Massachusetts, June [Online]. Available at: www.nber.org/system/files/working_papers/w25939/w25939.pdf (accessed: 19 January 2021).

Tooze, A. (2020a) 'Welcome to the final battle for the climate'. *Foreign Policy,* 17 October [Online]. Available at: https://foreignpolicy.com/2020/10/17/great-power-competition-climate-china-europe-japan/ (accessed: 17 October 2020).

———. (2020b) 'Carbon pricing and the exit from fossil fuels'. *Social Europe,* 6 July [Online]. Available at: www.socialeurope.eu/carbon-pricing-and-the-exit-from-fossil-fuels (accessed: 29 January 2021).

UN (United Nations). (2020) '2020 is a pivotal year for climate – UN Chief and COP26 President', 9 March [Online]. Available at: https://unfccc.int/news/2020-is-a-pivotal-year-for-climate-un-chief-and-cop26-president (accessed: 30 September 2020).

Wagner, G. and Weitzman, M. (2015) *Climate Shock: The Economic Consequences of a Hotter Planet.* Princeton: Princeton University Press.

Wolff, G.B. (2020) 'Europe should promote a Climate Club after the US elections'. Breugel.org, 10 December [Online]. Available at: www.bruegel.org/2020/12/europe-should-promote-a-climate-club-after-the-us-elections/ (accessed: 17 December 2020).

World Bank. (2020) Carbon pricing dashboard [Online]. Available at: https://carbonpricingdashboard.worldbank.org/ (accessed: 29 January 2021).

4

DEMOGRAPHICS, THE CHANGING INVESTMENT NARRATIVE LANDSCAPE, AND MARKET INCENTIVES

> An economic narrative is a contagious story that has the potential to change how people make economic decisions.
>
> *Shiller, 2019: 3*

> As with every major turning point in history, there comes a time for less talk and more action. That time is now. We can't quite imagine what's coming. We just know it's time. We can feel it. Crisis has sparked a collective demand for change.
>
> *State Street, 2020: 3*

> When climate change risks are considered material just as any other financial risk, there should be disclosures included. .... If they (company boards) don't think Paris-compliant assumptions are relevant for the valuing of assets, then it's important that that is known.
>
> *Carney, 2020b*

The urgent transition to net zero – the decarbonization of the economy – is not only a matter of ambitious government guardrails, glidepaths, and goals. Getting to a sustainable, resilient, more equitable and prosperous 2050 and beyond requires market narratives, and climate change decision making and strategies, shift sector to sector, company to company.

Shiller (2019) tells us that economic narratives – the stories investors, advisors, and employees tell one another about climate change – are pivotal drivers for economies and actors. These narratives inform us and help us determine what decisions we make and what conclusions we draw. And there has been a narrative shift. Evidence is mounting that investor and market narratives are reorienting towards the sustainability of the planet and net-zero goals. Investors and companies are

DOI: 10.4324/9781003037088-5

responding to customer demands for action on climate change and governance as they seek alternatives that responsibly and sustainably achieve their financial and investment goals. As Carney (2020b) explains, disclosure leverages and supports this narrative and investor shift. In addition, the coalescence of market narratives and investor demand around climate change goals can accelerate the transition to net zero.

This chapter presents evidence of shifting climate change narratives and actions in markets and firms and that the message of scientists and governments – that action is needed, as Greta Thunberg and Fatboy Slim would say, 'Right here, right now' (Thunberg, 2019; Fatboy Slim, 2019) – is beginning to be heard. As stories change, so do the decisions of millions of market participants and how they spend their trillions of dollars. It is potentially a game-changer for achieving our GHG emissions goals. When market narratives change, what was practically impossible becomes possible. This chapter underscores the importance of the narrative shift and the continuing, essential role of regulating, standard setting, and supervision that governments and regulators must play in setting expectations, altering incentives, and building upon the rate of transition.

The Gen X and millennial green wave

It is said that demography is destiny. This also applies to our thinking on climate crisis economics. One decade's baby bulge will be the next generation's decision makers and investors. The Gen Xers, born between 1965 and 1980 and 65 million strong in the US, and the millennial generation, born between 1981 and 1997 and 79 million strong in the US, are the market investors of today and the leaders and voters of the decades ahead. These younger workers, voters, and investors are alarmed by climate change. For example, American Gen X and millennial voters are more concerned than their parents about climate change, with 70 percent saying they worry about global warming (Gallup, 2018). A similar dynamic is seen in other countries. Thus, a 2019 Amnesty International poll across 22 countries found 41 percent of younger people cited global warming as the most important issue facing the world (Amnesty International, 2019). These young workers and voters are also more left of centre in the US and Europe (*The Atlantic*, 2019). Because of their sheer size and the timing of their coming of working age, their influence and the wealth transfer from their parents that is beginning to occur will cause a massive shift and surge in climate-related investment demands, underpinning and hastening the pace of the net-zero transition process. The new investor narrative is already reverberating across markets, advisory practices, and businesses.

Demand for investment strategies, options, and products that take climate and other social and governance issues into consideration, which come under the environmental, social, and governance (ESG) rubric, with a skew towards the 'E' for environmental, is rapidly rising. Sustainability investing and climate change concerns together signal a shift in the investment picture, one that alters the contours of markets' and firms' decisions. The 70-million-strong American baby boomer generation (born between 1946 and 1964) is beginning to transfer their wealth to their Gen X and millennial children and grandchildren just as we must collectively rush towards the zero-carbon goal in 2050. This process will be pivotal and accelerate the climate change market and economic transition in ways that governments alone cannot.

New investors demanding a greener future

Gen X and millennials in the advanced economies are on target to receive a vast transfer of wealth from their parents. Deloitte forecasts a transfer of US$24 trillion in assets (after taxes and charitable giving) from one generation to the other between 2015 and 2030, and that is likely an underestimate. Other estimates put the so-called great wealth transfer as high as US$68 trillion in America alone as estates change hands over the longer horizon between 2017 and 2061 (Cerulli Associates, 2018; Deloitte, 2015).

Similar asset transfer waves are underway and will be building across all advanced economies. This will accelerate the shift from a carbon economy to a decarbonized future. Money talks, and the cacophony of demands is unmistakable. As these demands for ESG investments mount, they are changing financial advisory, investment, and fund management businesses, reorienting firms to themselves demand more socially responsible, sustainable investments from the companies in which they invest. Polling sheds light on this.

A 2018 survey indicated that 87 percent of high-net-worth millennials considered a company's ESG track record important in their decision about whether to invest in it or not. It is not only the very wealthy young investors who are changing their focus to sustainability. Other polls find that 85 percent of millennials want to tailor their investments to their values and are looking at sustainable investing, and 50 percent are already beginning to shift their investment activity (Morgan Stanley, 2019). It is now widely recognized that this large and socially aware generation is not content with passive investments – i.e. personal investments that are divorced from company practices, behaviours, and strategies. The new, activist customers know what their principles are and want their investments to reflect them. As MSCI (2020) notes:

> Millennials, as well as women and, increasingly, individual investors of all ages and genders, are interested in directing their investments toward companies with good ESG records. This reflects a desire for their money not just to earn a return but to align with their personal values and contribute to the social good.
>
> *MSCI, 2020: 1*

Younger investors appear to be turning away from their parents' adherence to a market economy society in which the market is society (Carney, 2020) and towards a market, and investments, which instead must serve humanity and the planet. This difference in beliefs about the balance between markets and societal goals will have potentially lasting positive effects on markets, firms, and investment flows. It comes almost precisely at the time when we need markets to make the leap and help accelerate the transition. It is possible that these two generations might be about to do what their parents and grandparents failed to do: Act on their concern for the climate and our collective survival via millions of individual decisions on where and how to invest in the markets. State Street (2020), a leading advisory firm, believes the turning point on ESG investments for this growing cadre of committed young customers is finally here:

> With a greater emphasis on living according to their values, investors are increasingly ready to take a stand with their investment choices. We believe it is time for ESG investing to move from a check-the-box component of investment portfolios to a must-have ingredient in every portfolio.
>
> *State Street, 2020: 3*

This dynamic shift driven by young investors provides insight into their principles and values and should be a source of optimism. It turns out that rather than being only self-interested individuals focused exclusively on their devices – a common trope parroted by older workers about their younger colleagues – these young investors are opting for a new balance between markets and society and between economic value and planetary ecological value. The young are seeking a new definition of value that rejects a scorched earth, *laissez faire* market and depleted hothouse globe for one that instead rests more lightly upon a sustainable, resilient, inclusive foundation that conserves the planet for future generations. Many young people today appear to be rejecting the stark view of *homo economicus* in favour of *homo economicus sympatico*. This is a generational rejection of society as a market and an embracing of markets that must serve society, the planet, and future stability. I believe this mindset shift will continue and will result in a new balance between the economic utilitarian and the moral, ethical, and environmental. This is a positive, epochal change. It is in the process of altering markets meaningfully and permanently.

Shifting signals: Green is becoming synonymous with sound investments

Advisory and fund managers' incentives are shifting, driven by rising customer demand, and the words 'green' and 'investment' will increasingly become synonyms. Instead of marking-to-market, responsible global businesses will engage in marking-to-planet (Badre, 2020). Firms are beginning to benchmark, stress test, and risk manage their investments measuring against ESG and net-zero goals and to make decisions driven by those findings. Greenwashing – conveying a false impression or providing misleading information about how a company's products are environmentally sound – and virtue signalling are being replaced by directed, mission-driven investment decisions informed by the Paris Agreement goals and individual companies' plans to hit their own net-zero commitments. Firms will be measured and judged against their stewardship, net-zero plans, carbon footprints, new green products, and leadership in the transition. This massive, continuing investment shift will reallocate trillions of dollars towards 50 shades of green, and the sector leaders that stretch to reach the top will be rewarded. This will accelerate the transition. Profitability will follow and will be correlated to firms' performance against net-zero goals.

A green ESG wave no one can ignore

Expectations are changing and markets are assimilating an ESG narrative and shifting their focus and the stories they tell. Investment plans are being adjusted. Evidence shows that markets and investors are beginning to see ESG as a profitable and important part of any portfolio. There appears to be 'a recognition that sustainability [positively] influences risk and return' (BlackRock, 2020a). As active investors select the next Tesla, Scottish Power, or BrewDog to buy, others will demand easier, passive options that nonetheless impact the funds, the investment landscape, the markets, and firms' decisions and net-zero plans.

Indexing and exchange traded funds (ETFs) are now permitting a widening group of concerned long-term investors to place their money and 'mark to planet' – i.e. to factor in climate and sustainability in their decisions. These flows reached a record US$55 billion in 2019, amounting to US$220 billion in sustainable index ETFs in 2019. These flows did not reverse when the pandemic struck the globe. The early Covid-19 impact in March and April 2020 resulted in massive outflows from plain vanilla equity investments into the safety of cash and government bonds (even those paying negative rates). The pandemic did not result in similar outflows from ESG funds and ETFs. Instead, there was a modest inflow in investments and lower price volatility than the markets, or other ETFs (BlackRock, 2020b, 2020c, 2020d). Not only did ESG perform well in the pandemic, ESG ETFs outperformed the markets during the downturns and subsequent rallies. The signal from ESG investors was a long-term bet on direction. The performance was also not lost on market observers and seemed to send the message: Bet on investing in a (carefully selected) ESG climate-positive, net-zero portfolio if you are seeking medium- and long-term investments.

ESG investors are in it for the long term, and demand for passive ESG options will continue to surge. It is estimated that sustainable ETF investments will increase sixfold by 2030 to at least US$1.2 trillion (BlackRock, 2020a: 4). ESG investing is the green wave upon which markets will rise in the years ahead. BlackRock estimates that dedicated ESG funds in the US stood at US$350 billion in 2000 and had jumped to approximately US$850 billion in 2019 (BlackRock, 2020b: 4). Other estimates that use a looser definition of sustainability screening put the total at as high as US$16.5 trillion in the US in 2020, reporting a compound annual growth rate for sustainable investments of 14 percent (USSIF, 2020), or as high as US$26 trillion.

Figures for sustainable investing in Europe show similar jumps. Across all ESG strategies – which include sustainability-themed investment, best-in-class investment, exclusion of holdings from the investment universe, norms-based screening, ESG integration factors in financial analysis, engagement and voting on sustainability matters, and impact investing – flows continued to mount, with a competitive compound annual growth rate of 27 percent from 2015 to 2017. Funds under management with an ESG integration strategy grew from 2.6 trillion euros in 2015 to 4.2 trillion euros in 2017 (Eurosif, 2018: 16).

Just how massive the ESG shift is depends on how broadly or narrowly you define investment types.

Some advisory firms may just be rebranding and moving on – i.e. boosting the size of ESG funds without changing their composition. Such activity should become untenable in the medium term as disclosure will increasingly shed light on which firms are leading, which are stalling, and which are polluters. When investors discover that advisors are hoodwinking them, they will invest their funds elsewhere.

BlackRock (2020b) describes sustainability as the tectonic shift transforming investing and argues that:

> The direct impacts of climate change and the coming capital reallocation will reshape economic fundamentals, expected returns, and assessments of risk. Strategic asset allocations need to incorporate these implications in ways that go far beyond simply screening out certain stocks or securities.
>
> *BlackRock, 2020b: 2*

Clearly, today, the climate change ESG wave is visible to fund managers and institutional and individual investors. Long-term investors must consider ESG factors in allocation decisions and invest in companies that embrace the transition to a net-zero future. The investment management community is listening, and their narrative is shifting.

Firms need to get on board and set net-zero goals

The investment management community is demanding change from the companies they invest in. For example, Climate Action 100+, a lobby group whose members represent global investors with a collective US\$47 trillion in assets, announced in 2020 that it would begin judging 161 of the largest companies, collectively responsible for up to 80 percent of global industrial GHG emissions, by their progress towards net-zero carbon emissions (Climate Action 100+, n.d.).

Public and private investors from pension funds (such as CalPERS and the New York Pension Fund) to sovereign wealth funds (such as Singapore's Temasek) to asset managers (such as BlackRock) to international lending institutions (such as the EIB) are altering their investment policies and goals to plan for the carbon-neutral future (see Boxes A4.1.1–A4.1.5 in Appendix 4.1). These firm-level decisions have feedback-loop effects, causing still others to consider their climate change and net-zero positions and investment policies.

Finance and the net-zero transition

The EIB example shows how quasi-public-sector finance can assist. Banks and the finance sector also have an essential ongoing role in insuring, financing, and speeding the transition. Firms' decisions on who to insure, lend to, and advise and where to disinvest, what to back, and what to reject are powerful mechanisms that can pull forward our net-zero goals. As with other sectors, we can learn from the leaders (see Boxes A4.2.1–A4.2.4 in Appendix 4.2) and follow their examples, not only because it is the ethical, environmental, and socially right choice but also because it is the smart economic choice for the medium and long term.

The examples of Munich Re, AXA, BNP Paribas, and UBS provided in Appendix 4.2 underscore that the business and narrative shift is already underway, as is a widespread investment policy shift. This signalling by such giants in their markets, followed up with actions, investments, disinvestments, board resolutions, and shifting business strategies, alters our investment stories and cascades into decisions in the markets and inside firms.

Statements backed by the actions of millions of investors and scores of the world's largest asset managers in response to their clients are meaningful. When the statements are transformed into investment and disinvestment decisions, markets and firms pay attention. This is what real change looks like on the ground, in the markets, and is the beginning of narratives made real through action.

These signals from asset managers, insurers, and banks are increasing and unambiguous. In the future, most investment firms will make an increasing proportion of their investment decisions based in part on a firm's progress towards net zero. This will change market narratives and pay dividends. Investors will pay for sustainability and penalize polluters. As asset managers continue this shift, and more and more investors apply sustainability considerations, green firms will rise in value and laggards will fall. The alignment of companies with their own and the country's net-zero goals will become a material fiduciary factor in each company's success and impact its ability to secure investment and investor support.

I foresee that company boards of major firms will face suits by shareholders for failing to plan for and avoid climate change risks that materially affect firm profitability and sustainability and, by extension, shareholder returns. As UK board members are personally responsible for the risks they authorize, such actions will further clarify minds in the City of London and echo across boardrooms elsewhere.

Business leaders are waking up to the net-zero requirement and the rising ESG wave

Leading businesses and their CEOs understand the link between commitment to decarbonization and profitability. CEOs want their companies to succeed. They want their companies to reflect the concerns of their new customers, who are ESG focused, and of their young employees, who increasingly worry about climate change and who want to work for companies engaged in doing something about it. That requires a net-zero commitment and implementation plan. All this necessitates that CEOs shift strategies, set net-zero targets, and create opportunities out of the creative destruction in the markets.

CEOs and the leading net-zero firms are being rewarded. Investment flows towards their equity; they will seize market share and rise with the swelling green investment wave. Those firms that only pay lip service or that actively invest in sectors damaging the planet will see rising investment premiums, falling equity values, and shareholders increasingly revolting at annual general meetings. The worst offenders, according to Carney (2019), will lose market share, see equity values shrink and investors exit, and ultimately face bankruptcy.

Figure 4.1 shows the performance of the top 20 percent compared to the bottom 20 percent of companies based on changes in carbon intensity. The signal is clear: Companies that pursue business goals and strategies in line with net-zero reductions and goals are outperforming others. Moreover, as markets and firms respond to customers, the clear divergence in returns between leaders and laggards – the performance spread – will widen.

A CEO in 2025 will be in a leadership role in 2030, a time when our carbon budget will be all but exhausted. Do those CEOs want to preside over the failure of their businesses, impoverishment of their employees, and a dangerous future for their children and grandchildren? Will they refuse to factor in climate risks and the climate change transition in their businesses? The answer is increasingly an

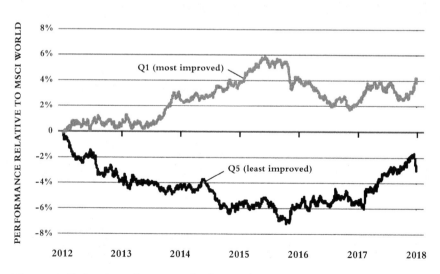

FIGURE 4.1 Carbon intensity matters for firm performance

Source: G30, 2020.

emphatic no. CEOs want to grasp the opportunity. That is what they are paid to do – seize opportunities. As the International Finance Corporation (IFC) puts it:

> Forward looking businesses are moving quickly to climate-smart investments because it is good for the bottom line. … investing in sustainability usually meets and often exceeds, the performance of comparable traditional investments.
>
> *IFC, 2020: 2*

There are huge opportunities to make money in the green economy. The IFC estimates there will be US$22 trillion in sustainable investment opportunities in the 21 leading emerging countries alone between 2016 and 2030. The Green Globalization 2.0 that must be constructed in the next three decades will present even larger investment opportunities.

Smart business leaders recognize the climate change and ESG shift and their responsibility to act on behalf of their stockholders, stakeholders, employees, and societies, and the many opportunities this presents. Their firms' long-term survival requires it, and investors and employees are demanding it. We can see that the returns are visible and will only go up. And company boards are demanding change. Polling of US public company directors shows corporate boards are under shareholder pressure to address climate risk as a key area of focus for their firms (NACD, 2018).

Shifting business strategies towards net-zero goals pays off

As we have seen, there are dynamic links between a firm's commitment to a net-zero transition, the alignment of a firm's business strategy, and returns. This can be visually represented in the green governance and returns matrix shown in Figure 4.2.

Firms in position A (perhaps Exxon or US coal giant Murray Energy) are not committed to the net-zero transition; i.e. the firms' business strategies are not aligned with climate change. The profits from this will create increasingly negative effects on the firms' returns, particularly once carbon is priced and regulated. Polluting firms will not go to the corporate graveyard easily; they will resist the transition and oppose action and regulation. As Tooze (2020) notes, there are several hundred large and profitable corporations whose businesses will be undermined by rapid and deep decarbonization. Exxon and others are on the list, but so are major Indian, Chinese, Australian, Russian, and Korean firms. These firms are fighting the climate change narrative shift, but they cannot resist the effects of climate reality and investor anger in the medium and long term. They will go bankrupt once investors understand their disregard for the planet, carbon is priced progressively, and transparency sheds ever greater light on their destructive actions.

Companies in position B have a greater degree of net-zero commitment, but their business strategy does not effectively reflect that commitment and they will

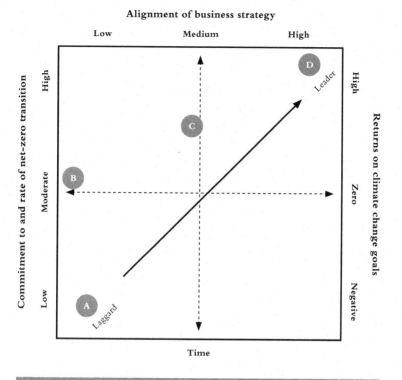

FIGURE 4.2 Green governance and returns matrix

therefore see modest returns. These companies are making changes, but too slowly, and will increasingly come under pressure as they are unfavourably compared against the leaders.

Companies in position C have a relatively high commitment to the transition and good alignment over time, and their returns are already apparent and will rise. These firms exhibit clearer leadership from the top and clear net-zero strategies that not only mitigate risks but also seize business opportunities, and this is reflected in their business models, plans, and products.

Companies in position D are first movers and leaders with clear plans and ongoing actions. Net-zero alignment of strategies, business lines, and products are embedded throughout the firm. Returns will consequently be greater. As first movers, costs may have been higher at the outset, but their position will strengthen as they consolidate their lead and reap the benefits from first-mover status and from new and existing growth opportunities.

Individual investors, CEOs, company boards, and employees need to decide which type of firm they want to engage with, invest in, lead, manage, and work for.

Ensuring the markets bend the GHG emissions curve towards net zero

Governments need to work to ensure the positive market narrative dynamics and decisions previously discussed translate into meaningful GHG emissions reductions and results. How can governments magnify the rate of change, extend the shift across all markets, and help investors determine who is delivering, what is working, and what is not? I divide the answer to this into three parts:

- *Disclosure.* Why making firms report their GHG plans and progress will speed change
- *Metrics.* Why designing oversight to ensure the metrics being used are clear, comparable, and understandable is important
- *New markets.* How supporting the creation and oversight of new markets that enlarge the real green investment opportunities available will speed the construction of Green Globalization 2.0.

Disclosure and transparency – what's not to like?

Disclosure standards for climate-related risks are crucial: Standardized disclosure allows investors to gauge the carbon intensity of an investment and act accordingly. Without such disclosure, investors cannot determine the risk and rewards. In 2021, climate risk disclosure is becoming a requirement for investors. Mark Carney, UN Special Envoy on Climate Action and Finance, has been a pivotal player in this change in disclosure norms.

In December 2015, Carney led the successful push for the creation of the Taskforce for Climate-Related Financial Disclosure (TCFD), under the oversight of the Financial Stability Board, which he then chaired. The TCFD issued a set of standards to the G20 leaders' forum and pressed businesses to apply the disclosure requirements. Carney was backed by allies in the finance and central banking community. The TCFD framework – issued in 2017 – was designed to pressure the world's largest firms to assess and report on the carbon intensity of their businesses. The framework's parameters are presented in Box 4.1.

BOX 4.1 TCFD REPORTING REQUIREMENTS

The TCFD is a business-led, technocrat-supported, relatively new part of the global architecture. It includes business leaders, policymakers, and supervisor and central bank support and input. Adhering to TCFD requires the following types of climate risk reporting:

- *Board level.* On oversight of climate risks and opportunities, including short-, medium-, and long-term goals
- *Senior management level.* On decision making and oversight of the business strategy financial planning associated with climate change risks and opportunities
- *Assessment.* Of the resilience of the organization's strategy given different climate scenarios, including Paris Agreement goals
- *Risk management.* On the processes for identifying, assessing, and managing climate-related risks, along with targets and performance against targets
- *Monitoring.* On risk management processes applied to climate change risk
- *Disclosure.* Of the metrics used by the organization to assess climate-related risks and address the organization's processes for managing climate-related risks.

Firms that agree to TCFD disclosure apply it to Scope 1, Scope 2, and, if appropriate, Scope 3 (GHG) emissions, and the related risks.

Scope 1. Direct GHG emissions from company-owned and controlled resources. These include:

- Stationary combustion and mobile combustion (i.e. all vehicles owned or controlled by a firm)
- Fugitive emissions leaks from GHGs (e.g. refrigeration, air conditioning units)
- Process emissions released during industrial processes and on-site manufacturing (e.g. production of CO_2 during cement manufacturing, factory fumes, chemicals).

Scope 2. Indirect emissions from the generation of purchased energy from a utility provider, i.e. all GHG emissions released into the atmosphere from the consumption of purchased electricity, steam, heating, and cooling.

Scope 3. Indirect emissions that are not owned but that occur in the value chain of the reporting company, including both upstream and downstream emissions, i.e. emissions that are linked to the company's operations.

Source: www.tcfd.org.

The TCFD framework is one of Carney's many strategic policy masterstrokes. The transparency request is eminently reasonable and potentially powerful. Firms would agree to disclose the carbon intensity of their businesses, investments, and supply chain. Carney was not, in 2015, asking firms to change their behaviour (although, of course, that goal was implicit); he merely wanted to provide investors with the information needed for them to make their own decisions based on the TCFD framework and annual reporting. CEOs might complain about the burden of reporting, but it is hard for them to oppose reporting on their businesses so that stakeholders and shareholders (the bosses in shareholder capitalism) can make decisions on the CEO's performance and competence vis-à-vis climate change risk and carbon intensity. Carney understood that making firms look at their own investments and business strategies and report on carbon intensity publicly could speed change and create positive feedback loops for the planet. That is what is now happening, four years on.

By 2020, the TCFD framework was being applied in over 1,440 of the world's largest companies, collectively worth over US$12.6 trillion dollars. Today, many of the world's largest companies are disclosing their businesses' carbon intensity. In 2021, TCFD is becoming a bare minimum investor question: Do you comply with TCFD? If not, why not? This creates pressure to act. It makes CEOs who commit to reporting engage in self-analysis and ask: Why are we doing business with firm X, in sector Y, with such a high carbon intensity; how are we advising them to change their practices; and if they are not going to change, what are we going to do about it? Moreover, TCFD reporting often ends up facilitating a company's own shift towards explicit net-zero goals and timelines. Here, again, we see self-reinforcing dynamics externally and internally. The effect on managers and businesses of the act of reporting and of considering GHG consequences and carbon intensity can begin to help shift the internal climate change narrative and business story in the firm among leaders and employees. Firms that agree to TCFD reporting are more likely to reappraise not only their own investments but also how they advise clients, borrowers, and customers on what they should do to begin to reappraise their own climate and GHG impact.

When I speak to chairs and CEOs, the most forward-looking are committed to net-zero goals, and they demonstrate that they are changing their firm's strategies, reporting against TCFD norms, disinvesting from polluting industries, and creating new products and commercial opportunities. You can tell whether a CEO is leading or just mouthing platitudes. Real leaders energize their firms and drive their employees. You can sense when firms are led well and are committed. This greening of corporate culture and narrative shift alters how business is done, which business is done, what is rejected, how risks are assessed, and how employees' performance is assessed. Taken together, leadership on net zero and thus TCFD translates into a meaningful series of changes for businesses that pay dividends for the firm and its employees.

Yet in 2021, too many firms still fail to apply the TCFD framework. Governments must therefore increase pressure on the laggards and shine a light on their polluting business choices. We cannot continue to have many firms and some entire major markets avoiding reporting against common carbon-intensity standards.

Make TCFD mandatory across the globe

How can the positive dynamics in TCFD reporting and reactions be markedly strengthened? This can be done by making TCFD reporting mandatory for all publicly traded firms and private firms over a certain size (G30, 2020). Governments should therefore make TCFD disclosure and public reporting required by law and national regulation (as some countries, such as France and New Zealand, already do).

As Carney (2020a) states: 'We think the time is now for mandatory disclosure'. Making TCFD mandatory would be a step change for all markets and expose polluters to the full glare of investors now looking to gauge whether firms are or are not aligning with society's climate change goals. Mandatory reporting should become the expected and demanded standard in all markets. Taking this simple and reasonable step will force all public companies to begin the urgent process of strategy realignment. Firms will be pushed to consider their transition plans and align their goals with planetary needs. They can opt to do nothing or to do too little, but this will have consequences. As Carney (2020) notes, applying TCFD means:

> Recognizing that when climate change risks are considered material that, just as any other financial risk, there should be disclosures included. …. If they (company boards) don't think Paris-compliant assumptions are relevant for the valuing of assets, then it's important that that is known.

Will making TCFD mandatory work? France provides a real-time example (see Box 4.2).

BOX 4.2 MAKING TCFD MANDATORY: THE FRENCH LESSON

In 2016, after signing the Paris Agreement, the French government made TCFD disclosure mandatory. French law required institutional investors (i.e. insurers, pension funds, and asset management firms), but not banks, to report annually on both their climate-related exposure and climate change mitigation policy. This provided a real-time experiment, namely: How would the investment decisions differ between those firms that were required to disclose and those that still could avoid doing so? Recent Banque de France research looked at this difference in investment.

Using a unique dataset of security-level portfolio holdings by each institutional sector in each euro area country, the researchers compared the portfolio choices of French institutional investors with those of French banks and all financial institutions located in other euro area countries. They found that investors subject to the new disclosure requirements curtailed their financing of fossil energy companies by some 40 percent compared to investors in the control group.

The lesson and message are clear: Make TCFD disclosure mandatory and accelerate the transition, pull forward decisions, shift investment outcomes, punish polluters, and reduce GHG emissions.

The UK recently signalled it will follow suit. Other countries, such as New Zealand, have also made TCFD mandatory. COP26 should announce all countries will apply TCFD disclosure standards by a set date, such as 2023.

Rapidly expanded and enhanced mandatory disclosure of climate change risks and opportunities will trigger positive feedback loops across sectors and amplify the effect of disclosure on our common GHG emissions. Mandating TCFD will alter incentives for firms through reporting. This coupled to shifting investor signals will increasingly push firms to change strategies if they have not already begun to do so. Five years from now, such TCFD disclosure will almost undoubtedly be the bare minimum as standards rise and investors respond to the information. For this process to work smoothly and efficiently we do need clearer standards and metrics against which one firm can be compared with another. Without that, we cannot be sure the reporting is real and meaningful.

Metrics matter for the economy and the planet

CEOs and firms complain they need common metrics to measure progress, to determine what is greenest, green, or brown and what to invest in and what to avoid. CEOs and firms want comparable, consistent standards. Neither they nor we (as consumers and investors) have time or the skillset to wade through hundreds of different standards, certifications, models, measures, and perhaps bogus stamps of approval. Companies want greater clarity on metrics. Global standard setters, and national authorities acting in coordination, need to respond. This means that national and global regulators need to update their standards and advisory rules to include sector-specific norms and expectations on climate change risk reporting metrics and processes and ensure the following:

- The Financial Stability Board must act as coordinating nexus.
- The Basle Committee on Banking Supervision must address climate change, physical, transition, and other risks within their remit and begin to lay out norms and a bank regulatory architecture that includes climate risk. The committee should do so drawing on the input of the NGFS.
- Securities regulators operating through the International Organization of Securities Commissions must consider disclosure requirements for listed firms

aimed at the consumer to ensure disclosure is meaningful, understandable, and consistent across markets.

- Insurance supervisors, operating via the International Association of Insurance Supervisors, must coordinate the updating of advisory rules on climate change risks for the sector.
- Accountancy authorities must coordinate to ensure that accounting and auditing standards and regulations include climate risk reporting. Once TCFD is mandatory, accountancy and audit firms will have a key role in determining whether the reporting is proper and correct.

Such detailed metrics matter.

In 2021, all the above-mentioned standard-setting bodies have begun this process, and these workstreams should gradually help clarify what is and is not acceptable as metrics and reporting on climate change risk. But there is a long way to go and we do not have much time, even though the technocrats are engaged. COP26 and G20 leaders should push the standard setters to move fast and report progress, with deliverables required by 2023. We cannot wait years and years for clarity and transparency across firms and markets. Committed CEOs and firms want this standard-setting engagement and direction, and they want clarity and predictability. For example, investor groups representing more than US$103 trillion in assets have demanded companies and auditors follow guidance from the International Accounting Standards Board (IASB) released in 2019. The IASB directions underscored factoring climate risks into company accounts, is already required within the existing rules, if relevant and material – even though most companies have yet to do so. The investor groups, including the signatories to the UN-backed Principles for Responsible Investment, have said companies need to explain the key assumptions made with regard to climate risk and make sure they are compatible with the goals of the Paris Climate Agreement – i.e. to start adhering to TCFD standards that promulgate the net-zero goals.

Central banks, as supervisors and regulators, must play key roles in supporting the shift in market incentives and expectations (see Box 4.3). Most have barely begun this task. I expect that over the next five-plus years this process – the promotion of climate change supervision, stress testing, and changes to risk and supervision assumptions – will accelerate.

BOX 4.3 THE ROLE OF CENTRAL BANKS IN SHIFTING INCENTIVES

Central banks will pay an increasingly important role as financial firms and markets are pressed to shift their incentives and alter - and long-term expectations and sentiments. Supervisory oversight is already changing. The Bank of England, for example, was the first major central bank to announce stress testing against climate change risk for insurers and banks.

> Central banks should stress test all regulated firms against climate change risk and make this a normal part of oversight by 2023 (G30, 2020).
>
> Supervisory climate risk requirements will become standard across markets. Central bank supervisors will progressively alter their assumptions and regulatory advice. The NGFS has been a leader is this community narrative leap on climate change.
>
> Going forward, central banks will increasingly influence business strategies among the world's largest banks as they add their regulatory voices to public and political demands for climate change action. Ultimately, I expect central banks will shift capital requirements for carbon-intensive lending – i.e. make lending to polluters more expensive to change incentives and market sentiments. This capital and regulatory shift will come, and faster in economies and markets where the net-zero goal is required by law.
>
> Importantly the US Federal Reserve System has finally recognized the risks that are mounting as a result of climate change. Supervisory changes and signalling by the US central bank will force US firms to reappraise and adjust their own assumptions and business strategies.
>
> The ECB is engaged in a review of climate risk supervision. I expect they will move quickly in 2021 and beyond to bring the power of the world's largest central bank to bear in support of the climate change transition.

As a result of these pressures, I expect – when TCFD becomes mandatory and firms are being supervised and stress tested against climate change risks – that when firms are audited, the audit will include a GHG audit and assessment against TCFD reporting and other supervision requirements. Boards and senior managers will want an external judgement on whether their climate risk assessments are correct, net-zero goals are being met, and their metrics are working; and if not, what they should do about it. This will be good for GHG outcomes and good for the planet. Technocrats and regulators acting together must further clarify the metrics framework for reporting and comparability. Investors, firms, finance, and consumers can then do much of the heavy lifting by pursuing their own goals and making their own choices, to mark-to-planet, and in doing so increasingly support the transition that is underway.

New markets and opportunities – parental oversight required

As markets, firms, actors, and investors get behind Green Globalization 2.0, new markets and opportunities will arise. Governments and their technocrats have a further important and ongoing task related to harnessing markets and speeding decarbonization: They must support but also oversee and, as necessary, regulate, new green markets. Governments must ensure we can seize the opportunities such new markets provide while avoiding dead-ends, scams, and instruments that pay the investor and creator in the short term but do nothing to achieve our GHG emissions goals.

Chapter 3 highlighted how poorly regulated ETSs can fail to deliver. So, too, can badly designed offset markets (such as the failed UN Clean Development Mechanism). Yet both types of markets are needed and must add measurably to GHG emissions goals. As I have stressed, the private sector left to its own devices cannot ensure good governance and planetary outcomes, so national governments must be willing to step in and oversee and supervise these markets.

The urgent need for reliable offset markets

The offset markets are a case in point. At present other than the EU ETS and California we lack effective regulated offset markets. I can buy offsets (and I do, every month), but I have no way of knowing if those GHG offsets are additive and are resulting in new reductions in GHG, or if I am just paying for worthless certification. This cannot be permitted to continue and is especially urgent because many firms want to buy offsets to achieve their net-zero short- and medium-term goals. Airlines want to buy offsets and charge their customers for them. Those offsets need to be created and to be real, measurable, audited, and overseen.

COP26 leaders have tasked Bill Winters, chief executive of Standard Chartered and former co-head of JPMorgan Chase's investment bank, and attorney Annette Nazareth, former commissioner of the US Securities and Exchange Commission, with leading a private sector task force to create offset markets that work and deliver for the planet. This is a formidable task. The market design is to be released in 2021 in advance of COP26. I hope they can deliver, but I remain to be convinced. Ultimately, I would prefer proper supervisory oversight once the markets are up and have begun to operate. I hope Winters and Nazareth will create the market and proper supervision, by national authorities, coordinating via existing mechanisms.

Vigilance against abuse and misuse

As these green markets and instruments are designed, proposed, and overseen, the question uppermost in the supervisors' and regulators' minds should be: Does security A, innovation B, or derivative C measurably support and accelerate GHG emissions reductions or not? If the answer is demonstrably yes, then the markets should be supported and the innovations scaled up. This should be the case with effectively regulated offset markets. If, however, a suggested innovation does not perform a useful net-zero purpose, it should be blocked or shut down. The planet does not have the time or the GHG budget to see finance and markets engaged in self-dealing, in destructive innovation that provides no planetary upside and only diversion of resources to enrich the few. Unfortunately, finance has a long history of creating markets and instruments that are corrosive, damaging, and destabilizing.

As green globalization takes hold, let us not be naïve. There will be efforts by immoral scoundrels to fleece investors. Securities and market regulators should take a tough line. They should work to ensure we do not waste time and planetary resources on foolish schemes with no net-zero benefit. A recent, true whale's tale illuminates both the huge unexamined value of nonhuman species to GHG reductions and the disappointing levels financial scam artists will sink to in their willingness to design environmental products with no value (see Box 4.4).

As new markets and specialisms take form around the net-zero-goal market transformation, governments, supervisors, regulators, and investors must be vigilant against scams and products that promise environmental benefits but do not deliver. It is essential that regulators take a tough and uncompromising stance towards scam

---

BOX 4.4 A REAL WHALE'S TALE

Recent work on the value of whales as carbon sinks, and as intrinsically magnificent beings, demonstrates how little we know until scientists or economists with an interdisciplinary mindset investigate. Chalmi *et al.* (2019) analysed whales and found that 'when it comes to saving the planet one whale is worth thousands of trees'.[a] Chalmi argues that whales should be considered an international public good because of the colossal amounts of carbon each whale sequesters during its lifetime of swimming the ocean's depths, consuming huge quantities of plankton. Chalmi *et al.* estimate the value of each whale at US$2 million.

Taken together, all whale species' carbon capture (at current – far too low – prices for carbon) is worth US$1 trillion to humanity. In 2021, there are 1.3 million whales in existence. If they were fully protected and the populations returned to preindustrial levels, whales would sequester 1.7 billion tons of CO_2 per year, more than the annual carbon emissions of Brazil.

Chalmi *et al.* are careful to note that they do not mean to imply that whales should be valued only for their carbon capture. Rather, this work was an attempt to make the whales' protection and defence clearer to economists and policymakers, some of whom, in Japan and Iceland, appear to view whales just as large fish to be slaughtered and eaten.

Financial scam artists took note of this new discovery and approached Chalmi, proposing to sell carbon-offset coins linked to whales as carbon sinks. The financiers do not own whales. Nor did they have any proposed means to increase whale populations. This was simply an exercise in selling a worthless financial product to people who feel good about whales, a sad testament to distorted innovation and unethical behaviour.

Note: [a] Chalmi *et al.*, 2019: 34.

artists and morally bankrupt operators who seek to make money out of the climate crisis without delivering GHG reductions.

Emerging markets and green investments

Advanced economy governments, institutions, technocrats, and their firms must support the deepening and extension of the markets for green assets and investments within emerging and lower-income economies. At present, investors looking for green opportunities in these regions can come up empty-handed. There simply are not enough investable opportunities. I recall the CEO of one of the world's largest investment funds lamenting the lack of green investment opportunities on the scale they require and need in those markets. Here, governments and their proxies (institutional and technocratic) have key galvanizing roles to play.

Technocrats should advise on emerging green market creation and oversight, and institutions such as the Bank of England, the Federal Reserve Board, and the ECB should second staff to support and improve the skillsets and the understanding and oversight of new green financial markets and products. Just as the EIB is pushing the pace of green investment and the transition in Europe, other regional multilateral development banks (MDBs) should have their resources increased dramatically and adjustments made in their mandates and investment policy statements to support green investments. MDBs should also be directed by state shareholders not to invest in and hold the investments but instead to sell them and seek new, green investments.

Here, again, the role of governments, technocrats, and the private sector is important. It is the collective effect of all three, acting on different aspects of market creation, support, and oversight, that can achieve the desired result. GHG reductions can stand on the back of deeper, well-supervised markets. More investable green opportunities mean good economic outcomes for emerging and lower-income countries.

An evolving market, properly overseen, will draw us towards our net-zero goal

It is time for another confession. When I conceived of this book and began my work, I thought it would be a much bleaker, doom-laden exercise than it has thankfully turned out to be. This discussion of climate change narrative shifts in markets that are being driven by demographics, investor sentiments, and market participant expectations and actions is positive, not negative. There are indications that a new market story and narrative on climate change, economic returns, and the future are taking hold. This is not going to reverse; rather, it will only build. This necessary and urgent narrative shift and market process can be accelerated and amplified with the right signals from governments, but governments need to be bold *now*. The perceived political roadblocks to faster action on carbon prices and regulation of the transition are being pushed against. Leaders, when they gather for COP26, can decisively move them out of the way. Governments must agree mandatory TCFD climate risk disclosure in all markets and for all publicly listed and large private firms. We know this works and will speed the rate of transition. Regulators should require disclosure from the end of 2023. Doing so will force laggards to act and require reappraisals of business strategies and alterations in investment decisions, and so speed the move away from fossil fuels. Mandatory disclosure is like motherhood and apple pie. No decent person can be against it. So just do it.

Governments and supervisors need to do further urgent work to clarify investor standards and metrics and ensure the metrics are meaningful, transparent, and comparable. Investors need predictability, credibility, and certainty as they shift their investment and, in doing so, speed the transition.

Technocrats and standards setters need to ensure international rules and norms are consistently applied and positively impact GHG emission reductions. Standard setters need to encourage a race for the top, not to the bottom. As with the carbon pricing schemes, standard setters need to establish the market expectation that standards in all sectors will progressively increase over time as we approach 2050 and be adjusted according to the progress (or not) of GHG reductions. Investors and firms should understand that such standards will become more stringent over time.

The professions that assist and advise firms, from consultants to accountants to auditors, are already responding to this market and regulatory shift. They and many other professions must be further drawn into the transition process.

New markets will be created and must be overseen. Properly supervised, they will speed us in the transition journey. Supervisors know what to be alert to and what to avoid. As needed, new institutional forums and organizations will also need to be empowered to oversee new carbon and offset markets. This is not impossible. We know what needs to be done. Vigilance against market abusers and vandals must be maintained, but again there are examples upon which to draw.

We can see that markets are essential accelerants in the net-zero transition in pulling forward the rate of green transformation. However, markets cannot do it unsupervised and should not be left alone. That has never been the right approach in the past, and it is not what is needed as we construct Green Globalization 2.0 in the decades ahead. We need consistent, credible, effective, predictable market oversight to speed us on the journey of decarbonization, including institutional innovations that can help us achieve our goals.

Appendix 4.1

The examples provided in this appendix offer a glimpse of the accelerating market narrative shift and investment policy shift that is underway and illuminate how investors are sending signals that are changing businesses' strategies and aligning them with the net-zero goal and climate change narrative.

BOX A4.1.1 CALPERS GETS GREEN-ISH

CalPERS, one of the world's largest public pension funds (with US$326 billion in assets in 2017) is a signatory to the Climate Action 100+ commitment to net zero by 2050. The fund is pressing the companies it invests in to decarbonize.

This fund was the first US asset owner to join the UN-convened Net-Zero Asset Owner Alliance, which commits to achieving portfolio emissions in line with the 1.5-degree-Celsius target for the real economy.

The company, which is following a gradual process of disinvestment from fossil fuel investments, has exited coal investments. This process is ongoing, and the company remains subject to sustained pressure to speed up its disinvestment plan.

CalPERS has a sustainable investments programme, which informs investment decisions aimed at the highest-value climate-change-related risks and opportunities. They are targeting opportunities to invest in companies whose goals include emissions reductions.

CalPERS supports carbon pricing, risk reporting, and fiscal measures in California and federally to support the transition.

The firm committed to begin reporting TCFD disclosure requirements.

Sources: CalPERS, 2020; responsible-investor.com, 2019.

BOX A4.1.2 NY PENSION FUND DROPS FOSSIL FUELS

New York State's US$226 billion pension fund announced in 2020 that it would begin disinvesting in fossil fuel stocks over the next five years and sell its shares of companies that contribute to global warming by 2040. The fund had already moved to sell its stock in coal companies. The fund had resisted calls for such a shift, but in 2020 its investment manager saw the fiduciary writing on the wall and began the shift away from fossil fuels. Thomas DiNapoli, the fund's manager, stated:

> New York State's pension fund is at the leading edge of investors addressing climate risk, because investing for the low-carbon future is essential to protect the fund's long-term value ... [and] future ability to provide investment returns in light of the global consensus on climate change.
>
> *New York Times*, 2020

The disinvestment plan is the result of an agreement between Mr DiNapoli and state lawmakers who, spurred by an eight-year campaign by climate activists, had been poised to pass legislation requiring him to sell fossil fuel stocks. This case, like the case of CalPERS in Box A4.1.1, has been sped up by activists seeking to change the narrative and investment decisions of public funds with employee pensions.

The NY pension fund has concluded that energy companies that do not reshape themselves to part with oil and gas are poor long-term bets. The message is clear: Stranded assets are coming to a market near you. Get out, rebalance, and seek new opportunities. The fund has published a wide-ranging climate action plan.[a]

Note: [a] OSC, 2019.

BOX A4.1.3 TEMASEK'S GREEN LEADERSHIP

Temasek, Singapore's SD$306 billion Sovereign Wealth Fund, is committed to the net-zero transition. The fund has announced it will reduce net GHG emissions attributable to its portfolio to 7 million tons CO_2 equivalent by 2030. This represents a drop by 50 percent in estimated GHG emissions attributable to their portfolio in 2010, or a quarter of estimated GHG emissions in 2020.

The firm's climate risk analysis includes applying an internal carbon price, which will guide decisions on new investments in the decade ahead. The fund conducts a sustainability review and applies an ESG screen to all new investments.

Temasek's CEO, Dilhan Pillay, has said the fund is

> committed to building a more sustainable planet for ourselves, and for future generations. ... This has led us to accelerate our investments into low-emission and resource-efficient companies, including in the areas of energy, food, waste, water, mobility and urban development.
>
> Temasek, 2020

As a major investor, Temasek is pressing the firms in which it invests to apply TCFD standards and has announced its support for them. This puts pressure on firms to conduct the analyses and has positive climate ripple effects across markets and firms.

Temasek supports the TCFD process. The fund is a long-term investor taking a generational view and is committed to deliver net-zero emissions in its entire portfolio by 2050.

BOX A4.1.4 BLACKROCK'S FINK TAKES A STAND

BlackRock, one of the world's largest asset managers, is led by Larry Fink. Fink is convinced of the central importance of climate change for financial markets. In his annual letter in 2020, he made clear his belief that: 'The evidence on climate risk is compelling investors to reassess core assumptions about modern finance'.[a] He signalled that change was afoot in his firm and in markets more broadly and was driving a profound reassessment of risk and asset values. Because capital markets pull future risk forward, the firm expects to see changes in capital allocation more quickly than changes to the climate itself. This is exactly what we want to see occurring. As Fink has underscored:

> Climate change is almost invariably the top issue that clients around the world raise with BlackRock. ... They are seeking to understand both the physical risks associated with climate change as well as the ways that climate policy will impact prices, costs, and demand across the entire economy.
>
> BlackRock, 2020

The firm has committed to placing sustainability at the centre of its investment approach, including making sustainability integral to portfolio construction and risk management; exiting investments that present a high sustainability-related risk, such as coal producers; launching new investment products that screen fossil fuels; and strengthening the commitment to sustainability and transparency in investment stewardship.

Note: [a] BlackRock, 2020.

BOX A4.1.5 EIB'S NET-ZERO SUCCESSES

In 2019, the EIB announced an energy investment policy review, which concluded that:

> Financing and advice must be directed at investments that cut emissions and combat climate change. We must create the jobs and growth in the renewable energy and energy efficiency sectors that will ensure the transition leaves behind no part of our societies and no region of the world.
>
> EIB, 2019

The EIB was signalling an end to fossil fuel investments.

The EIB has already been investing in projects that reduced GHG emission by 8 million tons annually during 2013–2017. In the five years running up to the 2019 announcement, the bank had already committed 50 billion euros in renewable energy and grid investments. The bank's investments will provide 38,000 megawatts of renewable capacity in Europe, supplying 45 million households.[a]

The EIB is a key vehicle in the European Green Deal and net-zero strategy, an essential actor in the financial markets, and a multiplier of public policy leverage and effect. Other global and regional MDBs should follow its lead. The time to intertwine and embed climate change goals across the financial system using public institutions to catalyze and speed change is upon us.

Note: [a] EIB, 2019.

Appendix 4.2

BOX A4.2.1 MUNICH RE AND CLIMATE RISK

If any business sector understands climate risk, it is the insurance and reinsurance sector. Understanding, predicting, and costing tail risk is essential to successful insurance and reinsurance. Munich Re is one of the world's largest, most dynamic reinsurers. The firm has been modelling climate risk for decades and is constantly adjusting its models depending on the latest scientific data and findings on floods, cyclones, storms, wildfires, droughts, and other natural phenomena. When you make an insurance claim for a severe weather event, Munich Re will ultimately insure the insurer against those losses.

The firm's leadership is committed to the net-zero transition. Change must always start inside a firm, with its own practices. The firm has achieved a 44 percent reduction in GHG emissions per employee since 2009. In 2019, the firm used renewables for 90 percent of its electricity use.

Munich Re will not invest in shares or bonds of a company if it derives more that 30 percent of its revenue from coal. It does not insure new coal-fired power stations or coal mines in advanced economies, and it has joined the Net-Zero Asset Owner Alliance[a] and has undertaken to transition its entire investment portfolio to net zero by 2050.

The firm has been carbon neutral since 2015 and adheres to the TCFD framework.

Note: [a] See www.unepfi.org/net-zero-alliance.

BOX A4.2.2 AXA AND THE CLIMATE TRANSITION

AXA, one of Europe's leading insurers, is engaged in the climate change transition. The firm is a member of the Net-Zero Asset Owner Alliance and has committed to reorienting its portfolio to achieve net-zero goals by 2050. AXA is divesting from carbon-intensive industries and will have divested from coal in the EU and the Organisation for Economic Co-operation and Development (OECD) by 2030 and in the world by 2040. The firm is in the process of doubling its green investments.

The firm is advising its clients on their climate risks and transition plans and is supporting the issuance of carbon transition bonds designed to help firms achieve climate change goals and strategies.

Internally, the firm has reduced its GHG emissions by 25 percent and has committed to source 100 percent of its electricity from renewables by 2025. AXA uses the TCFD framework and issues annual reports based on the framework.

BOX A4.2.3 BNP PARIBAS AND SUSTAINABLE FINANCE

BNP Paribas is France's leading bank in the rush to net zero. As BNP chairman Jean Lemierre states, banks are in the position 'to play a major role in creating an economy that is more respectful of the environment and more inclusive by choosing how to target [their] financing'.[a] The bank supports the French government's national net-zero goal. Mr Lemierre is clear that the firm must align its business with net-zero targets and is the principal advocate for the shift.

In 2020, the bank committed US$125 billion in support of the energy transition and sustainable development goals (SDGs). The bank ranks at the top for green finance and lending and is providing a plethora of green products to its customers: approximately 9 billion euros in credit for energy transition, energy efficiency loans, and sustainability-linked loans;[b] 10 billion euros in green bonds[c] (top-three performer); issuing the first SDGs bond; and 63 billion euros in socially responsible investment including 12 billion euros in green funds.

The bank is the first in the world to commit to a complete divestment from coal and stopped financing oil and gas fracking and reoriented towards renewable energy financing as a core market. The bank reports against the TCFD framework and has begun management of the bank's balance sheet against ESG criteria and the firm's climate change goals.

What matters is delivery. As BNP CEO Jean-Laurent Bonnafé stated: 'We need to be able to prove that we what we say is what we do'.[d]

Notes: [a] BNP, 2020: 2.
[b] Sustainability-linked loans are loans whose interest rate is linked to the achievement of environmental and social criteria matching the client's goals. BNP Paribas is ranked world leader in this new form of loan.
[c] The demand for green bonds is accelerating, and BNP Paribas was the third-largest issuer in the world in 2020.
[d] BNP Paribas, 2020: 10.

BOX A4.2.4 UBS AND CLIMATE CHANGE

UBS, Switzerland's leading investment bank, is leading on climate change goals. The bank's carbon-related asset exposure is low, at 0.8 percent of US$1.9 billion in assets, having been pushed down from 2.8 percent in 2017.

In contrast, climate-related sustainable investments stood at US$108 billion in 2019, up from US$87.5 billion in the prior year.

UBS is pressing for change with its oil and gas clients. In 2019, the firm voted for 44 climate-related resolutions.

UBS lending is aligning with climate goals. The firm will not provide project finance for coal plants globally and will only support financing for coal-fired firms if they have a transition plan that aligns with Paris Agreement net-zero goals. Nor will the firm finance offshore oil projects in the Arctic or greenfield oil sands projects.

Within the firm, the company is leading by example, reducing its GHG emissions by 71 percent in 2019 compared to 2004 levels.

UBS chairman Axel Weber understands what is at stake: 'If you look at the current research and you take some interest in that, you know this: climate change is for real. ... The speed at which the environment is changing is astonishing. So needs to be the speed at which we respond to that'.[a]

Note: [a] Yahoo Finance, 2019.

References

Amnesty International. (2019) 'Climate change ranks highest as vital issue of our time – Gen Z survey', 10 December [Online]. Available at: www.amnesty.org/en/latest/news/2019/12/climate-change-ranks-highest-as-vital-issue-of-our-time (accessed: 1 January 2021).

The Atlantic. (2019) 'European millennials are not like their American counterparts', 30 September [Online]. Available at: www.theatlantic.com/ideas/archive/2019/09/europes-young-not-so-woke/598783 (accessed: 1 January 2020).

Badre, B. (2020) *Voulons-nous (sérieusement) changer le monde?: Repenser le monde et la finance après le Covid-19*

BlackRock. (2020) 2019 Client Letter [Online]. Available at: www.blackrock.com/us/individual/blackrock-client-letter (accessed: December 2020).

———. (2020a) 'Reshaping sustainable investing' [Online]. Available at: www.ishares.com/us/literature/whitepaper/reshaping-sustainable-investing-en-us.pdf (accessed: 1 January 2021).

———. (2020b) 'Tectonic shift in sustainable investing'. Weekly Commentary, 30 March [Online]. Available at: www.blackrock.com/corporate/literature/market-commentary/weekly-investment-commentary-en-us-20200330-shift-to-sustainable-investing.pdf (accessed: 1 January 2020).

———. (2020c) 'Navigating uncertainty with ESG ETFs', 9 April [Online]. Available at: www.ishares.com/us/insights/etf-trends/navigating-uncertainty-with-esg-etfs (accessed: 1 January 2021).

———. (2020d) 'Sustainable resilience amid market volatility', April [Online]. Available at: www.ishares.com/uk/professional/en/literature/whitepaper/sustainable-resilience-en-emea-pc-brochure.pdf (accessed: 1 January 2021).

BNP Paribas. (2020) 'A leader in sustainable finance', June.

CalPERS. (2020) www.calpers.org.

Carney, M. (2019) 'Firms that ignore climate change will go bankrupt', *The Guardian*, 13 October [Online]. Available at: www.theguardian.com/environment/2019/oct/13/firms-ignoring-climate-crisis-bankrupt-mark-carney-bank-england-governor (accessed: 31 December 2020).

———. (2020) BBC4 Reith Lectures [Online]. Available at: www.bbc.co.uk/programmes/b00729d9 (accessed: 31 December 2020).

———. (2020a) 'We think the time is now for mandatory disclosure', 29 May [Online]. Available at: www.youtube.com/watch?v=D7_UPULzrgM (accessed: 1 January 2021).

———. (2020b) 'UN Envoy backs investors' call for increased climate risk disclosure'. *Insurance Journal*, 28 September [Online]. Available at: www.insurancejournal.com/news/international/2020/09/28/584194.htm (accessed: 18 January 2021).

Cerulli Associates. (2018) 'US high net worth and ultra-high net worth markets 2018' [Online]. Available at: https://info.cerulli.com/rs/960-BBE-213/images/HNW-2018-Pre-Release-Factsheet.pdf (accessed: 31 December 2020).

Chalmi, R., Cosimano, T., Fullenkamp, C., and Oztosun, S. (2019) 'Nature's solution to climate change'. *Finance and Development*, 56: 4 [Online]. Available at: www.imf.org/external/pubs/ft/fandd/2019/12/natures-solution-to-climate-change-chami.htm (accessed: 18 January 2021).

Climate Action 100+. (n.d.) 'How we got here' [Online]. Available at: www.climateaction100.org/approach/how-we-got-here (accessed: 1 January 2021).

Deloitte. (2015) '10 disruptive trends in wealth management' [Online]. Available at: www2.deloitte.com/content/dam/Deloitte/us/Documents/strategy/us-cons-disruptors-in-wealth-mgmt-final.pdf (accessed: 31 December 2020).

EIB (European Investment Bank). (2019) 'Big ambitions and investments for net-zero emissions', 14 April [Online]. Available at: www.eib.org/en/stories/energy-transformation (accessed: 30 December 2020).

Eurosif (European Sustainable Investment Forum). (2018) 'European SRI Study 2018' [Online]. Available at: www.eurosif.org/wp-content/uploads/2018/11/European-SRI-2018-Study-LR.pdf (accessed: 1 January 2020).

Fatboy Slim. (2019) A special mix by Fatboy Slim integrating Greta Thunberg's 'Right here. Right now' statement at the UN Climate Conference into his classic track of the same name [Online]. Available at: www.youtube.com/watch?v=bWvFcR7UtAI (accessed: 30 December 2020).

Gallup. (2018) 'Global warming age gap: Younger Americans most worried' [Online]. Available at: https://news.gallup.com/poll/234314/global-warming-age-gap-younger-americans-worried.aspx (accessed:1 January 2021).

G30 (Group of Thirty). (2020) 'Mainstreaming the transition to net zero' [Online]. Available at: https://group30.org/publications/detail/4791 (accessed: 31 December 2020).

IFC (International Finance Corporation). (2020) 'Climate investment opportunities in emerging markets' [Online]. Available at: www.ifc.org/wps/wcm/connect/59260145-ec2e-40de-97e6-3aa78b82b3c9/3503-IFC-Climate_Investment_Opportunity-Report-Dec-FINAL.pdf?MOD=AJPERES&CVID=lBLd6Xq (accessed: 29 January 2021).

Morgan Stanley. (2019) 'Sustainable signals' [Online]. Available at: www.morganstanley.com/content/dam/msdotcom/infographics/sustainable-investing/Sustainable_Signals_Individual_Investor_White_Paper_Final.pdf (accessed: 31 December 2020).

MSCI. (2020) 'Swipe to invest: The story behind millennial investing', March [Online]. Available at: www.msci.com/documents/10199/07e7a7d3-59c3-4d0b-b0b5-029e8fd3974b (accessed: 29 January 2021).

NACD (National Association of Corporate Directors). (2018) 'Building board climate competence to drive corporate climate performance', 12 June [Online]. Available at: https://blog.nacdonline.org/posts/climate-competence-performance (accessed: 1 January 2021).

New York Times. (2020) 'New York's $226 billion pension fund is dropping fossil fuel stocks', 9 December [Online]. Available at: www.nytimes.com/2020/12/09/nyregion/new-york-pension-fossil-fuels.html (accessed: 30 December 2020).

OSC (Office of the State Comptroller). (2019) 'Di Napoli releases climate action plan', 6 June [Online]. Available at: www.osc.state.ny.us/press/releases/2019/06/dinapoli-releases-climate-action-plan (accessed: 30 December 2020).

responsible-investor.com. (2019) 'CalPERS says it plans to align with TCFD amid new California climate legislation', 25 June [Online]. Available at: www.responsible-investor.com/articles/calpers-tcfd (accessed: 30 December 2020).

Shiller, R.J. (2019) *Narrative Economics: How Stories Go Viral and Drive Major Economic Events*. Princeton: Princeton University Press.

State Street. (2020) 'ESG: From tipping point to turning point'. White Paper, July [Online]. Available at: www.ssga.com/library-content/pdfs/etf/spdr-esg-investing-tipping-point-to-turning-point.pdf (accessed: 29 January 2021).

Temasek. (2020) 'Temasek review 2020: Committed, resilient, together', 8 September [Online]. Available at: www.temasek.com.sg/en/news-and-views/news-room/news/2020/temasek-review-2020-committed-resilient-together (accessed: 30 December 2020).

Thunberg, G. (2019) UN Speech by Greta Thunberg, 23 September [Online]. Available at: www.youtube.com/watch?v=QJo9uXn2QxU (accessed: 30 December 2020).

Tooze, A. (2020) 'Welcome to the final battle for the climate', 17 October [Online]. Available at: https://foreignpolicy.com/2020/10/17/great-power-competition-climate-china-europe-japan (accessed: 29 January 2021).

Yahoo Finance. (2019) [Online]. Available at: https://finance.yahoo.com/news/axel-weber-ubs-climate-change-global-warming-sustainable-investing-esg-105851987.html (accessed: 3 January 2020).

5

BUILDING A DECARBONIZED WORLD

Institutional innovations that reinforce market outcomes

> Ensuring carbon and other greenhouse gases are priced efficiently is a challenge. … Carbon Councils that embody the expertise, credibility, and predictability needed to supervise and oversee markets should be designed in order to ensure the delivery of real, positive planetary outcomes and dramatically lowered greenhouse gas emissions.
>
> *Rey, 2020*

Making our net-zero commitments stronger, quasi-binding, more resilient, more rules-based, and more consistent in their application and effect requires the creation of new institutions, the reform of existing ones, or both. Governments across the world must commit to net-zero goals and finally agree to price carbon and gradually raise prices to achieve net zero and internalise the cost of pollution and global warming across all markets and sectors. Leaders can implement taxes (as in Canada, Sweden, and elsewhere) and construct and reform ETSs (as in the US, EU, and elsewhere). Carbon prices should rise and converge to minimise distortions. However, all these green commitments and GHG policy mechanisms and goals can be undermined and weakened unless we have effective institutions and forums tasked with supervision, oversight, and monitoring compliance (Rey, 2020). National climate change action has a history of setting goals that are then unmatched by implementation and oversight. This is not surprising. If a short-term economic crisis occurs, governments and leaders, facing re-election, shift their focus, and long-term GHG goals can be missed as policymakers turn towards the short-term needs of the economy and away from the required simultaneous focus on a net-zero future.

DOI: 10.4324/9781003037088-6

Green institutions for a sustainable tomorrow

If we are to align policies across societies and economies to achieve decarbonization, we need institutions that pursue our common goals across election cycles outside the nakedly political process, once leaders take the decision to set net-zero goals. Now that a growing number of national governments are committed to net-zero goals, it is time to create the organizations to help change incentives and speed the transition and rate of decarbonization.

This chapter sets out institutional constructs that can help us secure our goals – principally the creation of a World Carbon Organization and the creation of National Carbon Banks. These institutions, international and national in scope, would become the supervisors and enforcers of our agreed GHG goals. Supervision, oversight, enforcement, and rules are needed to create certainty and credibility out of volatility and doubt.

A World Carbon Organization could referee disputes between and among countries over carbon pricing and its consistent application and adjudicate the application of the carbon border tax. National Carbon Banks can simultaneously pursue and help create sustained deflationary pressure on GHG emissions as economies and firms set net-zero goals and seek to achieve them. I have said previously that self-regulation is an oxymoron. On carbon and GHG emissions, this is demonstrably true. We have examples of how institutions can effectively oversee and shift markets and their behaviours. We need to apply organizational lessons from existing organizations and from central banking and supervision to the task of achieving net zero.

Self-reporting and monitoring without enforcement are not enough

At present, we have the UN COP process, countries' self-reporting of goals and progress, and national GHG reduction commitments. However, the self-monitoring and self-assessment of progress against national commitments is mixed and often understandably reflective of a country's position on climate change. Those states committed to the planetary common good and net-zero goals (such as the Netherlands, the UK, and Sweden) report in detail. States that are laggards produce reporting that is much less robust or subject to direct and troubling political manipulation (such as the US and Russia). In addition, there are many middle- and lower-income states that lack the funds and capacity to effectively design and implement their GHG goals, let alone report on them effectively.

In 2021, the review process is weak and unsatisfactory. At present, there is nothing that can be done when states fail to live up to their GHG reduction commitments. There is no global enforcement or options for dispute settlement over carbon pricing. I am hopeful that at COP26 or soon thereafter many more (a majority of?) states will begin this essential process of committing to shifting carbon pricing and incentives. When this occurs, we will need to design an institution to oversee pricing and to establish, embed, and defend the new international carbon pricing rules and norms.

We should create a World Carbon Organization

It is naïve to assume that all governments and sectors will play by the rules and act in good faith and that all sectors will be charged equivalent pricing for carbon without some form of international enforcement mechanism. Therefore, an important step in global coordination, cooperation, compliance, and enforcement would be the creation of the World Carbon Organization (WCO). The WCO would help ensure consistency and compatibility across national GHG carbon pricing regimes. It would ensure that carbon pricing rules are applied fairly, carbon dumping across borders is stopped, and there is a consistent approach towards carbon pricing. The WCO would be the forum for dispute settlement over the consistent application and implementation of national and international GHG pricing, offsets, and markets. The WCO, with a dispute settlement procedure modelled on the WTO approach, could make determinations when disputes arose between or among countries over the application of carbon prices in comparable sectors in disparate ways.

Replicate WTO success in the carbon pricing space

The WTO, although repeatedly attacked and undermined by President Trump throughout his term in office, remains the most important, widely accepted forum for trade dispute settlement. Its birth heralded an expansion of free and fair global trade, and it has served as a forum for discussion and agreement on the extension of common trade rules. It also generates pressure on non-members to join, while providing enforcement of common rules and approaches.

The WTO's quasi-judicial dispute settlement procedure, which adjudicates disputes over trade rule application and adherence, is widely recognized as fair and workable. Since the WTO's creation, there have been far fewer bilateral disputes and fewer rule breaches, while hundreds of disputes have been successfully and pro-ductively dealt with within the multilateral processes. On trade and trade rules, the WTO provides predictability and certainty, two key attributes that governments, businesses, and investors look for as they plan their strategies for growth.

The WCO would, once established, help replicate and provide such predict-ability and certainty on carbon pricing and trading and would aid international carbon pricing coordination and oversight. It could adjudicate carbon border tax-ation rules and the application of dumping rules, oversee carbon offset trading norms and practices.

The WCO could adjudicate and establish norms

The WCO is needed if, as is likely, a carbon border tariff is applied to discourage free riders and the weakest link phenomenon, where some countries consistently under deliver thus undermining the entire GHG emissions goal. The WCO can exclude and punish free riders by applying the carbon border tariff. The latter may occur if a country with carbon pricing commitments which is inside the coalition of the willing fails to apply pricing fairly and stringently; the WCO could determine when this is the case and ensure consistency across countries carbon pricing. Without a WCO, carbon border tariffs would be bilateral, uncoordinated, and disputatious, degenerating into damaging clashes between those applying carbon prices and free riders. The WCO would also stand between those inside the carbon pricing club, as an institutional, rules-based bulwark against bullying by those outside the climate change consensus. Carbon intensity and GHG pollution divergences already exist across many markets and sectors and will likely persist without an international mechanism to adjudicate and to penalize bad actors who are avoiding paying for their pollution.

Take the steel sector. Steel produced in the US and Mexico uses much less GHG-intensive manufacturing processes than in China. If this GHG intensity divergence persists after the differential application of national carbon pricing to Chinese steel, should this carbon dumping be allowed or should it be penalized with border taxes when Chinese steel is shipped to North America? I would strongly urge the latter. Ideally, we need a WCO dispute mechanism that can be triggered in the event of carbon dumping. We should replicate the WTO's success in addressing disputes through a balanced forum that ensures fair carbon pricing outcomes. The number of such disputes will likely be large in the decades ahead as carbon pricing is applied, dodged, and disputed. This can perhaps be handled initially without a WCO via the proposed unilateral carbon border adjustment tax, but we could ultimately see a large number of bilateral actions, sector by sector and firm by firm. In the medium to long term, if we want to sustain net-zero momentum, decarbonize, and reform the supply chains of global trade while helping to ensure transparency and minimal friction, WCO oversight and adjudication is the better option.

This problem is analogous to the 'spaghetti bowl' of bilateral deals that confounds businesses in the absence of WTO agreements on norms. It can work, but it is very confusing as firms have difficulty knowing which rules apply to which countries and to which goods. Rather than create a new, green variety of bilateral carbon pricing spaghetti, leaders should ultimately opt for a multilateral deal on pricing enforcement rules and norms.

The WCO would over time establish those norms and adjudicate based on agreed norms and rules, which would help establish certainty and predictability. It would avoid power-based diplomacy and judgements. The WCO could also advocate for carbon solutions, act as a repository for knowledge and dissemination of best decarbonization regulatory practices and metrics, and help ensure equity in the treatment of lower-income countries, with burden sharing and carve-outs for the poorest states.

A WCO would become a sister to the WTO, with which it would work in parallel to oversee global trade and ensure it is conducted in line with international agreements on carbon pricing and the net-zero glidepath. Much as the WTO has ensured a rules-based international trading system and order, the creation of a WCO dispute settlement process would tie together the state parties administering the global and national carbon pricing systems and help build consistency, convergence, clarity, and fairness.

Critics may ask: Why create a new institution? We already have the UN, the IPCC, and the COP process. That is understandable. Creating institutions is not always easy and is not always the solution, but we currently have patchy carbon pricing mechanisms and no agreed way to ensure constancy, comparability, and enforcement. If we are to decarbonize world trade and do so in an equitable and coordinated manner, and one that speeds the transition while enforcing rules and norms, we need the WCO to operate alongside the purely diplomatic and consensus-led institutions.

We have seen that institutional creation can indeed be galvanizing, as in the case of the NGFS, a new forum created in 2017 that today leads the central banking policy discussion and debate on climate change and net zero. The NGFS has increased the rate of policy convergence, brought together leaders, sped up policy action, and coordinated with other international and national institutions and standard-setting bodies. The WCO could potentially achieve similar success. Importantly, we have also seen that the UN is no place for market oversight mechanisms and processes; that is not its strong suit. Yet we must have an institution that can oversee global carbon markets, which will grow rapidly, to help us collectively reduce GHG emissions.

Monitoring new markets

The WCO could also perform a key role in new market oversight and coordination. The WCO could, for example, monitor the emerging, new offset markets that are being created and that will become so important to our climate change goals. The WCO in the first instance could promulgate best practices for offset programmes, disseminate data, and act as a forum for coordination and cooperation. Ultimately, a WCO should make recommendations as to the operation and oversight of the new and reformed carbon markets, much as other standard-setting bodies already do in the fields of banking, insurance, and markets, coordinated via the Financial Stability Board. The WCO's overarching role would be to oversee global, regional, and national carbon pricing regimes and provide a framework within which agreed pricing could be reviewed, coordinated, and adjusted.

The WCO would also be the global forum for technical, expert-level discussion of carbon pricing, trading, and taxation. The WCO could become an intellectual repository, an incubator of specialist carbon measurement, metrics, and analysis. It could help nurture skillsets and provide intellectual resources to help make the global and national carbon markets work more smoothly and press for an upward convergence on prices, practices, and supervision. The nascent carbon supervisory community, an emerging epistemic community of carbon pricing expertise, needs a home. The WCO would be the place.

While the WCO, working alongside the WTO, could help ensure consistency, comparability, and fairness in carbon pricing internationally, it could not ensure net-zero goals are being achieved and implemented nationally. To help achieve that, each country should establish its own National Carbon Bank or equivalent organization.

Why we need National Carbon Banks

National Carbon Banks (NCBs) are needed to oversee the implementation of national carbon prices and markets in each jurisdiction. Governments must agree to achieve net-zero goals by 2050. Governments should then pass on the oversight of the implementation of the process to an NCB (or similar council or institution) to carry out the day-to-day supervision and technical oversight of achievement of those GHG goals (G30, 2020). Support for such a move is gradually building, with Mark Carney and Janet Yellen, President Biden's treasury secretary, on record as supporting rising carbon prices and the creation of Carbon Councils to make policy effective (Bloomberg, 2020). How and whether these policy mechanisms are taken up in the US will have a significant impact.[1]

What should the NCB mandate be? It should be tasked with monitoring national net-zero goals in the short, medium, and long term and with supporting the goal by overseeing and making recommendations on the regulation of carbon pricing and carbon markets and ensuring proper supervision.

Several countries already have this type of technocratic institution, as presented in Box 5.1.

BOX 5.1 COUNTRIES WITH CLIMATE CHANGE TECHNOCRATIC INSTITUTIONS: THE EXAMPLES OF THE UK, IRELAND, SWEDEN, AND FRANCE

The UK's Committee on Climate Change

In 2008, the UK established an independent, statutory Committee on Climate Change (CCC), tasked with advising the UK government on emissions targets. The CCC is comprised of former ministers, senior civil servants, academics from a mix of backgrounds, and experts from the energy industry and produces an annual report on progress towards meeting these targets. The government is required to explain divergences from goals and targets.

The CCC assesses long-term emission reduction targets and recommends specific five-year carbon budgets. These intermediate targets provide transparency about the trajectory towards net zero and help hold the government to account.

The CCC is a cornerstone of climate policy in the UK, notwithstanding the fact that the UK government did not grant the CCC executive powers beyond issuing public recommendations. The CCC has nonetheless success-fully influenced the climate policy of seven UK governments.

CCC recommendations have changed government GHG targets, policies, and laws. All the CCC five-year target budgets have been accepted by the government and are expected to be met. The CCC's recommendations for the 2023–2027 and 2028–2032 budgets have been adopted by the UK govern-ment. The current agreed budget calls for emissions to be reduced to 43 per-cent of their 1990 levels by 2032.

The CCC advised the UK government to increase the level of its climate ambitions and agree a net-zero target. The UK has as a result formally agreed to achieve carbon neutrality by 2050.

These successes do not mean that the CCC is exclusively laudatory of the UK government's GHG performance. The CCC has repeatedly criticized the government's plan to bank any emission reductions in excess of what was required by previous carbon budgets and offset them against future emissions. This is important in 2021 because the UK is currently expected to overshoot the recommended carbon budget during both 2023–2027 and 2028–2032 and wants to use its overperformance in prior years to offset this failure.

Ireland's Climate Change Advisory Council

In 2015, Ireland passed the Climate Action and Low Carbon Development Act, which created the Climate Change Advisory Council. The council began operating in 2016.

Ireland's Climate Change Advisory Council is an independent advisory body tasked with assessing and advising on how Ireland can achieve the transition to a low carbon, climate-resilient, environmentally sustainable economy.

The council conducts evidence-based analysis on how best to respond to the impact of climate change and provides the government with advice on the most effective policies to assist with Ireland's transition.

The council provides regular reports (periodic and annual) regarding Ireland's progress in achieving its national policy goals and the GHG targets agreed by the EU and is empowered to speak broadly on climate change goals and progress (CCAC, 2020).

Sweden's Climate Policy Council

In 2017, Sweden's parliament made the country's goal of net zero by 2045 (and carbon negative thereafter) legally binding. This is a stretch goal, leaving most other states (such as the US[a]) far behind. The parliament created the Climate Policy Council as an independent scientific council with the task of assessing whether the government is achieving its stated climate goals. The Climate Policy Council reviews sector goals and performance against the goals on an ongoing basis, reporting annually, and provides critical analysis of government performance against GHG goals. Specifically, the council:

- Evaluates whether policy regarding different areas contributes to or counteracts the climate goals
- Reviews the effects of existing and planned policies from a broad societal perspective
- Identifies policy areas where additional measures are needed to achieve climate goals.

Importantly, the council is also tasked with evaluating the analytical methods and models that are the basis for the policy and with contributing to the debate regarding climate policy. The public can view the country's GHG performance using Panorama, a web-based visualization tool of progress against the goals (Climate Council [Sweden], 2020).

The French Haute Conseil pour le Climat

In 2018, France established the Haute Conseil pour le Climat (High Council on Climate, HCC), a panel of independent experts. Its role is to assess the French government's climate policies, and after mandating in 2019 the legal requirement to achieve net zero by 2050, the HCC began holding the government to account against its own stated goal.

The HCC has already criticized the government for missing its earlier GHG goals and for doing too little to change consumer behaviour. In the HCC's first annual report, the panel highlighted France's insufficient efforts in reducing GHG emissions in relation to international commitments and called on Paris to radically change its climate policy. The HCC warned that, 'The current pace of economic transformation is insufficient because transition, efficiency and energy saving policies are not at the heart of public action'. The HCC called on the government to set tougher policies to achieve its goals. The HCC annual report concluded that France was a good student 'with bad results' and needs to work much harder.[b]

At this early stage, the French construct appears less robust than the British CCC, but the HCC signalled in its first annual report that it will not shirk its reporting duties. The HCC is making clear it must be taken seriously and its criticism must be addressed.

Notes: [a] Swedish per capita GHG emissions will be at 1 ton by 2045. In contrast, US GHG emissions per capita in 2015 stood at 15.5 tons.
[b] Euractiv, 2019.

Creating independent national oversight organizations in every country, as strong or stronger than the UK CCC, would materially help translate national climate change goals agreed at the political level into monitored deliverables over the short, medium, and long term.

As of 2021, France, Ireland, New Zealand,[2] and Sweden have created their own councils modelled on the UK CCC.

The UK Climate Change Act created the destination but, as Quartz observes, the CCC sets the milestones. It is the kind of approach that other countries need to adopt:

> Just like central banks set monetary policy with economic stability in mind – regardless of which political party is in power – the CCC independently sets out its vision for a carbon policy with climate resilience in mind. ... This thorough and evidence-based approach gives the CCC great credibility ... the UK government often listens.
>
> *Quartz, 2019*

Creating strong, independent NCBs in every country would have many benefits

Creating an NCB removes from politicians the difficult and unpopular task of making hard choices and recommendations on carbon pricing, policy oversight, and enforcement once a goal is set and legislated. Few politicians are interested in or want to be seen planning to increase prices or determining supplies of offsets or emissions allowances or the steepness of the glidepath year to year. These decisions are too politically contentious. It is far better to leave supervision and oversight to a technocratic body you can blame and castigate. We can see this working in central banking.

During his term, US President Trump often used Jay Powell, Chair of the US Federal Reserve System (the US central bank), as a tweet target whenever he felt a monetary policy stance was unpopular or wrong. President Trump could blame Powell and castigate him, and he did. Yet, in 2020, the president claimed credit for the results of an expansionary monetary policy and its stimulatory effects before and during the Covid-19 pandemic response.

Or look back further to the 1970s, to then-Federal Reserve Chair Paul Volcker, who famously hiked interest rates into the high teens to crush runaway inflation (Volcker, 2019). President Carter knew this was necessary, as did others, yet the move was hugely unpopular. President Carter could deny responsibility and much of the blame fell on Mr Volcker. He bore it like the giant he was, with stoicism and strength. Just as we leave central bankers to take the heat for many tough decisions today, politicians should leave the tough carbon markets decisions to independent NCBs.

Creating an NCB also avoids the danger of politically motivated abuse. We saw how under-pressure politicians provided far too many emissions permits in the early EU ETS and other schemes, massively deflating the price of carbon, under-mining what could be called the desired 'sound carbon policy', and distorting the carbon market. An NCB could ensure such abuses are avoided by removing the opportunity for political gain via price and market manipulation. We all benefit from technocratic oversight of carbon pricing and markets (European Commission, 2020).[3]

An NCB should pursue a transparent and predictable policy pathway, learning from central bank operations and history. An NCB could thus issue regular reports on GHG targets, emissions pricing, permits, and regulation, sector to sector. Markets, investors, and interested individuals could use these communications to make decisions about consumption, investing, and the economic outlook, just as we do today when we monitor the Bank of England or the Federal Reserve monetary policy AQ: announcements and reporting.

Just as central banks have jettisoned the opacity of Greenspan for the transparency of Bernanke, so should NCBs do likewise in their oversight of carbon markets and pricing pathways. Central banks today provide forward guidance so investors and businesses can plan for months and sometimes years ahead with some general sense of the glidepath of interest rates. Policy is adjusted with reference to shifting data and indicators of the health of the economy over time, such as with reference to publicly understood inflation and employment targets.

Similarly, NCBs should lay out the various criteria they use for carbon pricing and market decisions, the factors they consider, and the impact of changes in GHG stock and flow, tipping points, and scientific findings and developments. Such transparency would allow markets to anticipate, shift, and bring forward decisions based on expected or forecast NCB actions. We can see this happening now in the carbon price in the EU ETS; markets are anticipating shifts and adjusting pricing upwards, changing expectations accordingly. We also saw similar positive effects of predictability and transparency in the US CO_2 market. Over time, NCBs could similarly build predictability and therefore enhance their credibility, increasing their market impact.

As NCB credibility solidified, this would bring forward market decisions on net zero and lower the cost of overall decarbonization. What do I mean? If an NCB is credible and predictable in its policies and communication, it will support carbon market operations and trading. Investors and analysts will understand the policy pathway ahead and can plan accordingly. In this way, NCB credibility pays policy dividends and can thus help create a self-reinforcing strengthening of the desired carbon policy pricing goals.

Look again at the case of inflation in the US in the 1970s and at Paul Volcker. He re-established the credibility of the US central bank by taking hard decisions and telegraphing this difficult monetary policy tightening repeatedly to the markets. Markets then began to assume a lower inflation rate was to be expected, was possible, and would be sustained and defended going forwards. The decades of low inflation since the 1970s can be traced directly to Volcker's re-establishment of central bank credibility in the fight against inflation. Chair Volcker's credibility amplified the effect of his monetary policy statements and actions, as well as all those central bankers who came after him and learned the credibility lesson.

A demonstration of the importance of credibility was also in evidence in the famous statement by Mario Draghi, then-president of the European Central Bank, when he stated he would 'do whatever it takes' to stabilize the Eurozone during the 2011 crisis. Draghi and the ECB had credibility, so markets responded accordingly. Investors viewed the ECB and Draghi as credible, and they assumed he would stand by his statement. This amplified the effect of his policy statement across markets and European economies, and in the end his words had impact without action, as investors calmed in response to the strong statement. The ECB case (and others) demonstrates that credibility matters for policy effectiveness, especially in crises (Bems *et al.*, 2018).

NCB credibility would operate in the same fashion. Markets, investors, businesses, and individuals would come to understand and believe NCB statements about actions aimed at achieving carbon pricing goals, anticipate adjustments to GHG prices and emissions markets, and make changes to their plans and actions. I have stressed the importance of harnessing markets to speed the transition to net zero so as to maximize the benefits that might accrue from the green industrial revolution that is underway and speeding up. A predictable and credible NCB would strengthen carbon markets, harness market dynamics, and smooth our transition pathway.

What of crisis response? Here, too, an NCB, once established and operating with effective communication, predictability, and credibility, could *in extremis* take sudden drastic action if required, faster than politicians and insulated from political influence. Consider the following possible scenario: NCBs are established in all states and are overseeing carbon markets and pricing based on national net-zero pathways and GHG emissions. Scientists discover that frozen methane hydrates are thawing at a rapid and increasing rate and large volumes of gas are being released into the atmosphere (such scientific evidence indeed emerged in 2020). NCBs could act and raise the price of carbon, coordinating action across borders to keep countries on target. In the meantime, and in parallel, governments could begin to take other complementary policy actions to adjust to the exogenous shock of possible massive methane releases. My point here is that in a crisis, NCBs could act faster in a coordinated fashion on pricing. They would not need to wait for a renegotiation of the National Parties' Commitments or the next fraught COP meeting, hampered by intransigent national leaders unwilling to take further, possibly costly, action.

A central bank parallel can also be seen in the Covid-19 pandemic response. Central banks responded incredibly fast, providing in early March 2020 unlimited US dollar liquidity when the world's economy was shut down. Swap lines were immediately triggered. Leading central banks sent similar signals and took similar actions. Monetary policies were eased across the board in days, not weeks or months.

Independent NCBs could (and should) act in an emergency even when carbon policy decisions might be economically painful in the short term, just as Paul

Volcker did in the 1970s. NCBs would not need to wait and to parse the political and electoral implications. Finally, and importantly, an independent NCB can act with alacrity and strength when market abuses are found. Just as central banks and market regulators intervene when supervisors detect abuses and failures at the market or firm level, so too should the NCB.

An NCB with carbon market supervisory authority could help ensure markets were not abused and manipulated. If that did occur, corrective actions could take place. At present, this type of micro- or macro-prudential oversight of carbon markets is largely missing (except in the EU ETS). NCBs could fill that gap and support ethical and efficient market operations. For instance, if an NCB discovered a major firm cooking their carbon TCFD reporting, or gaming and abusing the markets, or distorting pricing (as we have seen occurring in Europe and in UN offsets markets), the NCB should be empowered to intervene and fine abusers. An NCB should in extreme cases also be able to withdraw a firm's licence to emit GHG if severe breaches of market carbon conduct and behaviour were found.

The importance of independence

The UK CCC example (see Box 5.1) demonstrates the importance of independence for effective operation of an NCB or carbon council. Getting to net zero requires wide-ranging and ongoing changes in policymaking and implementation across many levels and the application and oversight of a rising carbon price even if this is politically unpopular, especially in the short- to medium-term as societies and markets adjust. This underscores the importance of executive independence for NCBs. NCBs must be politically unbiased and independent from political influence once their leadership is appointed. Such institutions need distance from the political process so they can carry out their mandates unmolested by short-term political pressures and the demands of fossil fuel lobbyists.

NCB leaders must have executive independence so they retain the ability (and the authority) to respond when governments fail to achieve interim net-zero goals, when carbon pricing must be adjusted, or when markets need signalling that a policy change is required. We would come to expect them to do so, much as we do with the central bank governors of today, whose independence and credibility empowers them to deliver tough messages as required to stabilize economies in response to crises.

Accountability … of course

To balance this independence, NCB leaders should be confirmed by legislatures and required to regularly report to them. NCB leaders should also present their views and findings to the press and to the public at large in pursuit of their mission and climate change mandate, in furtherance of clear communication and transparency. Just as the governor of the Bank of England or the governor of the Bank of Japan must explain their policy decisions regularly and present to legislative committees, so should the members and chair of the NCB.

Institutional innovation in support of Green Globalization 2.0

I have stressed previously that the net-zero transition cannot be an exercise in incrementalism. Getting to net zero by 2050 requires a wholesale realignment of our policies and of governmental and business practices across many sectors and markets. This will require a radical redesign – a new, green industrial revolution – that our public institutional architecture needs to support. Some states and governments such as Sweden, the UK, and Denmark are broadly on target and have created mechanisms, policies, and institutions to facilitate the transition, but a great many other countries, such as the US and China, most notably, have yet to do so.

Crucially, these institutions, even if they are created, are not a substitute for a broad-based understanding of, discussion about, and support for the necessary climate change net-zero societal journey. In the end, if you create an institution in good faith and it lacks broad support and backing, it cannot do its job and it can be damaged or dismantled. Australia provides just such a sobering lesson (see Box 5.2).

BOX 5.2 AUSTRALIA'S TROUBLING EXAMPLE

Institutional innovation without a strong consensus is perilous. In such instances the creation of expert committees is not a substitute for a broad political consensus to address climate change. In 2011, the Australian government formed the Climate Commission as an independent entity to provide reliable and authoritative information about climate change on the continent. It was a good idea and a sound step, but the public discourse in the country did not at that time support the expert body or defend it from its fossil fuel enemies. Unfortunately, the commission was disbanded a mere two years later following the election of a new government, whose leadership disdained climate change facts despite rising temperatures and raging wildfires. While members of the disbanded commission soon set up a new Climate Council, it is an independent, nonprofit organization with no statutory underpinning.

The Australian Climate Council that now operates is strident in its criticism of the government's failure on stated GHG goals. This is to be applauded. For example, the council has excoriated the government, declaring, 'The Federal Government's own published data shows Australia's greenhouse gas pollution levels are expected to continue to go up over the next decade. Australia's emissions in 2030 are expected to be higher than today'.[a] The council is active and activist. However, the (un-official) council is not meaningfully impacting the government's policy debate, and yet, as we know, the world continues to warm, and the policy and public consensus can and does shift.

Note: [a] Climate Council (Australia), 2018.

This troubling example from Australia shows you cannot run too far ahead of the political consensus. Governments must work to change the consensus narrative and can do so, but they cannot create one out of nothing. However, it is also clear that a national climate change consensus can shift and can do so quite dramatically when conditions change in the face of reality.

Searing reality intervenes

The consensus among the Australian public has shifted dramatically since 2013. The horrific 2019 fires that raged down the South East coast of the country appear to have been a public opinion tipping point, as well as a climate tipping point. An electorate, historically resistant to climate change realities, is today reappraising climate change risk. By the autumn of 2020, concern about the impact of climate change hit a record high, with 80 percent of Australians polled believing the country was already experiencing problems caused by climate change and 83 percent supporting the closure of coal-fired power stations. This radical awakening was fuelled by wildfire. Half of the respondents said fossil fuel producers should pay for climate action and nearly three-quarters thought Australia should be a global leader in combatting climate change. In addition, three in five (59 percent) would prefer Australia's post-Covid-19 economic recovery to be powered by investment in renewable energy. Four in five South Australians (81 percent) think tackling climate change will create opportunities for new jobs and investment (Australia Institute, 2020). In summary, it is increasingly clear that Australia's electorate, confronted by their own climate tipping point, is making a leap to a new story and narrative about climate change and the country's role in addressing it.

Australian businesses are also waking up to climate risks and the need for action. For instance, in 2020, the Australian Energy Council committed to the net-zero goal by 2050 – a significant shift. The AEC stated, 'The first step to reducing carbon emissions is agreement on a long-term target which can act as the starting point for constructive consensus. ... Settling on an economy-wide target will let us then decide the best ways to get there and what policy and mechanisms could be applied' (AEC, 2020).

The recent Australian experience is both sobering and potentially ultimately positive. After all, it shows sudden shifts in public perceptions, narratives, and stories can occur, prompted by harsh climate change reality. Part of the answer to those who say we cannot achieve our net-zero goals – i.e. make the leap – because the public in country X or country Y is opposed is to retort that our individual and collective stories and understanding about climate change, and our place in addressing it, can and do shift, sometimes very suddenly indeed.

I believe now is the time for the Australian government to recognize the climate emergency and reauthorize the Climate Council and re-empower it to hold politicians to their public climate commitments. Australia should also legally mandate a net-zero goal for 2050. In 2021, the country's leaders are visibly lagging behind their populace and voters. Australia's political class needs to pay attention to their constituents' demands, begin the transition, and speed the industrial transformation that must come either in a planned, dynamic, economically additive manner, or else foolishly wait and be forced to shift by events at much greater economic and societal cost a few years from now.

Build the architecture of a green, decarbonized tomorrow

Creating a WCO or NCBs will be neither fast nor easy. These additions to the institutional architecture require delegation of regulatory power upwards (in the case of the WCO) and the devolution of regulatory powers downwards (to NCBs from governments). However, this type of organizational innovation is needed to coordinate and foster convergence on rising carbon prices and to support comparable and compatible regimes, while minimizing friction and tensions and ensuring fairness and at the same time punishing laggards and free riders. As countries work to create the contours of a green, decarbonized tomorrow, they need institutions that are fit for purpose and that take lessons from other forums and apply them to carbon markets pricing oversight and supervision. The names we use for the forums matter less than their functional effectiveness.

In 2021, we have clear gaps in the architecture and need to design workable solutions. As COP leaders and governments press ahead with agreements on rising prices for carbon, agreeing national glidepaths to net zero, and standing up and reforming carbon and offset markets, they need expert institutions and communities that can share the burden. We should start thinking now about the form these institutions should take.

As states seek to smooth the green transition, policymakers have a great deal to concern themselves with. Policymakers rightly focus on aligning and embedding net-zero goals, setting interim targets, and reporting across governments, the economy, and each sector. Delegating the difficult task of carbon pricing and market oversight to technocrats is the best, depoliticized additive option, since once goals are set, difficult pricing, dispute settlement, and market supervision decisions are not roles politicians should play. Leave that to others who can testify and report on progress as well as failures. We can see this works. Policymakers should learn the lessons from the WTO, the UK CCC, and the EU ETS. Leaders should then begin to design institutions to oversee carbon markets and their operation and leave the duty to oversee and supervise carbon prices and markets to expert technicians and technocratic communities with the necessary mandates, empowered to get the job done, internationally and nationally. Properly designed and empowered, a WCO and NCBs can support carbon pricing, net-zero goals, and market operation and oversight.

Politicians can then focus on the many remaining fiscal and other regulatory incentives and penalties needed to speed transition and decarbonization and support innovation, sector to sector. As countries construct the guardrails for the green markets of tomorrow and the green industrial revolution, governments must ensure that the transition begins swiftly across all countries, sectors, and industries to get to a Green Globalization 2.0 by 2050. Governments must steepen the green technology diffusion S-curve as they pursue decarbonization by 2050. Governments can then focus on the hard job of ensuring rapid green technology innovation and diffusion across all sectors – energy, transport, agriculture, construction, and industry. That is a huge and urgent job in and of itself. Governments should leave carbon market regulation to properly tasked and empowered international and national technocrats.

Notes

1 Secretary Yellen might opt to create a separate Carbon Council, or perhaps she might opt to embed responsibility within the Financial Stability Oversight Council, an existing interdepartmental forum. The latter option is not ideal. For an NCB or Carbon Council to be effective, it should be independent from political pressures and able to make recommendations without departmental haggling and horse trading. The NCB or Carbon Council needs to be separate from the political process.
2 See www.govt.nz/organisations/climate-change-commission.
3 The European Commission stepped in after discovering the extent of the market abuses, altering oversight, emissions permit numbers, and supervision. Those steps meaningfully shifted the operation of the EU ETS from a failure into a functioning rising market.

References

Australia Institute. (2020) 'Climate of the nation 2020: South Australians concerned about climate fires, want renewables led recovery' [Online]. Available at: www.tai.org.au/content/climate-nation-2020-south-australians-concerned-about-climate-fires-want-renewables-led (accessed: 30 October 2020).

AEC (Australian Energy Council). (2020) 'Australian Energy Council backs net zero emissions by 2050' [Online]. Available at: www.energycouncil.com.au/news/australian-energy-council-backs-net-zero-emissions-by-2050 (accessed: 30 October 2020).

Bems, R., Caselli, F., Grigoli, F., Gruss, B., and Lian. W. (2018) 'Central bank credibility pays off in times of stress'. IMFBlog, 3 October [Online]. Available at: https://blogs.imf.org/2018/10/03/central-bank-credibility-pays-off-in-times-of-stress/ (accessed: 4 January 2021).

Bloomberg. (2020) 'Yellen gets a shot to put treasury clout into climate fight', 11 December. [Online]. Available at: www.bloomberg.com/news/articles/2020-12-11/yellen-gets-a-shot-to-throw-treasury-s-clout-into-climate-fight (accessed: 4 January 2021).

CCAC (Climate Change Advisory Council). (2020) [Online]. Available at: www.climatecouncil.ie/aboutus (accessed: 4 January 2020).

Climate Council (Australia). (2018) 'Let's get something straight – Australia is not on target to meet its Paris climate target' [Online]. Available at: www.climatecouncil.org.au/australia-not-on-track-to-meet-climate-targets (accessed: 4 January 2021).

Climate Council (Sweden). (2020). 'The Swedish Climate Policy Council' [Online]. Available at: www.klimatpolitiskaradet.se/en/summary-in-english (accessed: 4 January 2021).

Euractiv. (2019) 'France is a "good student with bad results" when it comes to climate policy', 27 July [Online]. Available at: www.euractiv.com/section/energy-environment/news/france-is-a-good-student-with-bad-results-when-it-comes-to-climate-policy (accessed: 4 January 2021).

European Commission. (2020) 'Ensuring the integrity of the European carbon market' [Online]. Available at: https://ec.europa.eu/clima/policies/ets/oversight_en (accessed: 4 January 2020).

G30 (Group of Thirty). (2020) 'Mainstreaming the transition to a net zero economy' [Online]. Available at: www.g30.org (accessed: 4 January 2020).

Quartz. (2019) 'The UK's trailblazing advantage against climate change', 11 July [Online]. Available at: https://qz.com/1662800/the-uks-committee-on-climate-change-is-a-model-for-all-countries (accessed: 4 January 2021).

Rey, H. (2020) 'G30 calls for urgent and practical steps to speed transition to a net-zero economy', 8 October [Online]. Available at: www.prweb.com/releases/g30_calls_for_urgent_and_practical_steps_to_speed_transition_to_a_net_zero_economy/prweb17448548.htm (accessed: 20 January 2021).

Volcker, P. (2019) *Keeping At It*. Princeton: Princeton University Press.

6

A GREENING OF INDUSTRIAL POLICY

Speeding diffusion and the achievement of net zero

> Empirical studies of diffusion … [resemble] an S-curve: a slow period of early take-up is followed by a phase of rapid adoption and then a gradual approach to satiation. … The diffusion process is, in this view, analogous to the process by which epidemics spread: each user of the new technology passes information on to one or more non-users who, in turn, adopt the technology and spread the word.
>
> *Geroski, 1999*

Announcements of ambitious stretch net-zero targets are essential. Short-, medium-, and long-term plans for the progressive implementation, oversight, and monitoring of those plans are also required to convert policy statements into on-the-ground reality. Pricing carbon at a minimum level, with the price rising in a predictable and credible fashion over time, is key. So, too, are the institutional underpinnings to ensure compliance and enforcement by states operating within a globalized greening economy. Reorienting incentives through a combination of new narratives underpinning regulation, the spurring of business strategy shifts, and growing public pressure can assist, as can harnessing the dynamics of marking-to-planet in the net-zero transition. Yet, still more is needed to bend the GHG emissions curve country to country and globally. Today, governments must continue to support and speed up the net-zero technology diffusion (Geroski, 1999) and transition through all sectors, from utilities to transport, to construction, to agriculture, to airlines, to shipping, and beyond.

In 2021, no sector has fully decarbonized. While some sectors are more decarbonized than others, many have barely begun the necessary transition and must be pushed and supported as they do so. This requires a greening, a reseeding, of industrial policy. State power and authority must be used to speed the rate of change, sector to sector.

DOI: 10.4324/9781003037088-7

Whether these policies are described as climate transition planning, green innovation support, or green industrial policy does not matter. Governments should use the language and narratives that are most useful in the national and local context and with each audience. The key is the goal: to shift incentives and market dynamics sector to sector and speed the rate of industrial transformation, diffusion, and decarbonization. Governments, by placing sustainability at the centre of the policy process, can demonstrate net-zero goals. They can signal that progressively strengthened environmental standards are not an obstacle to a competitive economy and manufacturing sector but are the foundation of economic growth. Governments, especially laggard polluting giants such as the United States, need to abandon ideological opposition to industrial policy generally and green industrial policy specifically as a tool of societal and economic planning.

This is a war against carbon. We need to treat it as such. As Stiglitz (2019) has noted:

> When the US was attacked during the second world war no-one asked, 'Can we afford to fight the war?' It was an existential matter. We could not afford not to fight it. The same goes for the climate crisis.

The power and authority of states, their resources, and public agencies are key to fighting this war and to reorienting economic and regulatory incentives to national and global goals. A greening of governments' industrial policies can set the stage for a just, sustainable, resilient, and equitable economy. This will not happen spontaneously.

Public policy as enabler, supporter, and backer

Evidence shows that when governments commit resources, especially in the early stages, to nascent technologies through research and development (R&D), seed funds, and strategic innovation policies, it can pay off. Mazucatto (2015) has demonstrated that many of the dominant technologies of today resulted from publicly funded research. She disassembled an iPhone and identified which parts (most of them) relied on public funding in the first, pre-commercial period. Mazucatto's case is compelling but is dismissed by free-market fundamentalists because the conclusion does not fit with the myth of a genius CEO or a firm alone creating their own wealth and success.

Public support is, indeed, often essential at the beginning. An activist government role is needed, not to pick specific companies as winners or losers but to support the process of innovation at a time when firms do not, because the returns are at that point not yet obvious or because the venture capitalists and investors are backing other, easier bets.

Many regions and countries have begun an industrial and regulatory policy realignment. For example, the European Green Deal seeks to speed up the rate of change, aided by ongoing and expected shifts by the ECB and by national governments and regulators. The Biden green industrial policy will (US Congress permitting) funnel a stream of resources to clean tech R&D, to wind and solar technologies, and will begin to build the infrastructure for the economy of tomorrow and reset and reestablish regulatory guardrails and glidepaths. Analysts are forecasting sizeable multipliers and broad-based economic benefits from this investment surge, which will be accompanied by amplified shifts in private sector investments and markets.

As we saw, Sweden altered its taxation and regulatory stance on carbon decades ago without ill effect. The result was, instead, sustained GHG emissions reductions accompanied by strong economic growth. The country did not engage in picking individual winners and losers. The carbon tax's effects changed the balance and incentives. Firms and markets responded because their success or failure now depended in greater part on their ability and willingness to adapt to the carbon-costed normal. Setting a carbon tax and regulatory guardrails allowed markets and individuals to continue to make their own choices.

The Norwegian example of an aggressive EV adoption timeline and planning also illustrates that well-understood, transparent, consistent public policy is key to harnessing markets. The government decided on the net-zero goal; applied regulatory and industrial policy levers to support it; and provided incentives, penalties, and enforcement. The markets and individuals did the rest. Today, Norway has the highest percentage of EV adoption in the world because the government established an ambitious multiyear framework for the policy and for industry's transition.

The French example of remarkably rapid reforestation is also a triumph of public policy designed to address GHG emissions and deepen our carbon sinks. The French approach is activist, directional, clear in its goal, and effective. It doesn't select individual oaks to back one against another. Rather, it incentivizes landowners to grow more trees to better steward and care for the land. Today, a third of France is covered by forest, a green success story.

The key point is that a greening of industrial policies must accompany regulatory guardrails and limits. This is not insidious, creeping socialism. Such accusations are cover for Anglo-American neoliberal attacks on the role of the state, grounded in an outdated conception of what the state should and should not do to regulate the economy and foster environmental and planetary net-zero outcomes. Conservatives and libertarians on the political spectrum tell scare stories of government bureaucrats taking control over individuals' decision making. This is nonsense.

Governments already use industrial policy, and that is not a bad thing. All governments work to construct, explicitly and implicitly, economies and regulatory states that are a mirror of, and support the continuation of, national models. All government policies benefit some and negatively impact others.

In America, they just don't call it industrial policy. Rather, it is more obscure, insidious, and destructive of the common good. Opacity allows corporate interests that benefit from the luxury of bemoaning industrial policies to cash in on loopholes and tax treatments that have the effect of shifting incentives in their favour.

For example, US tax law allows real estate developers to roll over reported losses year to year and potentially avoid paying any tax. The *New York Times* (2020) reported that former President Trump paid only US$750 in taxes in 2017 and 2018, and that in 10 of the prior 15 years, he paid no taxes at all. Trump used a tax policy that favours commercial real estate over other productive investments. This is not called industrial policy, but it has the same effect, changing incentives and creating winners and losers.

Another example is the US tax treatment of private equity (PE) investments. Partners in PE firms borrow money to purchase firms. The purchased firm takes on the debt and PE partners manage the firm, charge high fees, and strip dividends and profits out of the now indebted company. PE partners are paid what is called 'carried interest', i.e. a share of any profits that the general partners of PE receive as compensation regardless of whether they contributed any initial funds to the purchase. This leverage-funded profit is taxed at 20 percent, below the normal tax rate for high wage earners of 37 percent in the US (hedge fund profits are also treated the same way in the US). This is a tax policy bias in favour of PE and hedge fund partners. It is a bias towards debt over equity and is distortionary in its effect on markets, capital allocation, and capital accumulation. Because of this tax policy, PE hedge fund firms are larger, more powerful, and more influential in the US. Here again, US policy shifts incentives and alters markets and investor decisions.

Consider also the failure by US authorities (until very recently) to apply antitrust and effective tax collection policies to the world's largest technology firms such as Amazon, Apple, Facebook, and Google. Their founders are today's tech equivalent of the steel and banking barons of the nineteenth century. The firms are immensely powerful globally and intrusive, monopolistic, dominant, and anticompetitive. Yet most of these firms pay almost no tax, despite their massive revenue streams, because of aggressive tax avoidance. The US government could reform tax policies and close loopholes, but it has not. Instead, US industrial policy is biased in favour of tech and is supportive of wealth concentration on an historic scale. Twenty-twenty-one is a gilded age if you live and work in Palo Alto, Seattle, or San Francisco, but not so much if you live in Uniontown, Pennsylvania, or Cleveland, Ohio.

Many countries pursue explicit industrial policies. German governments have for decades incentivized small, medium, and large industrial exporters, with tax and regulatory policies designed to foster industrial firms ahead of finance. Germany has also created a social and industrial model that supports labour and integrates it into firm decision making at the highest levels through workers' council representation. This has resulted in higher wages, lower inequality, and a highly skilled, educated, and productive workforce, from apprentices through to senior managers. Finance in the German economy is not divorced from the economy and its obligations to the state and the street, i.e. the people. Finance for speculative gain at the cost of society is viewed as vulture capitalism. The German model is not industrial policy and social democracy run amuck. It is an example of what can be achieved when industrial policy and the economy are made to work towards societal goals that benefit the many, not the few.

There are other national versions of activist industrial policy, such as China's state capitalist directed model, Korea's Chaebol model, Japan's mighty industrial model of the 1980s and 1990s, and Singapore's autocratic meritocratic model.

The point is that as governments accelerate us towards national net-zero goals, designed for their circumstances, economies, and societies, they must redesign their industrial policies to get from here to there by 2050.

This greening of industrial policy can support early-stage, pre-commercial innovation and R&D; recast regulations to reorient incentives; and accelerate the diffusion of new technologies, practices, and approaches that can decarbonize the economy. I would rewrite Bill Clinton's famous exhortation from the 1990s, 'It's the economy, stupid', as 'It's the diffusion rate, stupid'. What do I mean?

Diffusion rates, technology costs, innovation, and the net-zero transition

How fast a technology is developed and adopted is known as the diffusion rate. Research shows this rate follows an 'S-curve' – i.e. slow at the beginning of technology development and uptake, followed by a rapid steeping of the curve as the cost of the technology falls, feeding back into adoption rates, before the rate of diffusion flattens as a technology becomes mature and more fully used. Government industrial policy support for new technologies and their diffusion is essential because:

> Roughly half of the reductions that the world needs to swiftly achieve net-zero emissions in the coming decades must come from technologies that have not yet reached the market today.
>
> *Roberts, 2020*

Roberts adds that 'aggressive innovation will be required'. Our green net-zero future depends to a high degree on the design of government policies that speed the rate of new technology diffusion and innovation and the widespread adoption of new technologies that cut GHG emissions. Public policies and climate crisis economics as applied to diffusion should strive to:

- Steepen and bring the 'S' of the curve closer to us in time, i.e. hasten the rapid uptake of green technologies
- Use incentives, penalties, targets, phases-outs, and other levers to steepen the central section of the S-curve.

Steepening the curve and using levers can help trigger drops in the cost of – and a hastening of the scaling-up of – technologies, creating positive feedback loops in adoption, diffusion, pricing, and the achievement of net-zero goals.

Figure 6.1 depicts a diffusion price matrix, a stylized diffusion S-curve, with a price curve superimposed on it. The two interact in a dynamic process creating a

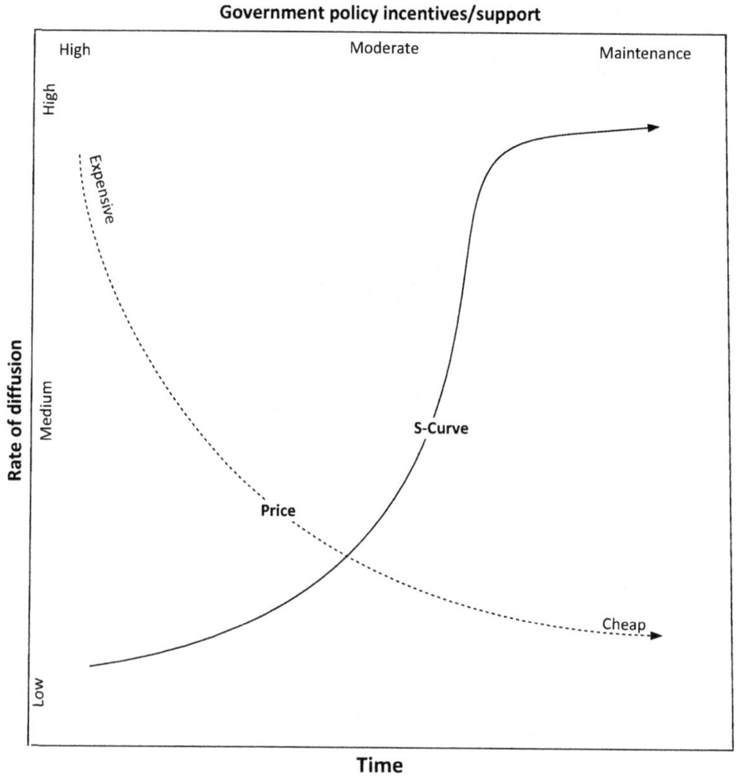

FIGURE 6.1 Diffusion price matrix

feedback loop. As a technology is adopted and diffusion speeds up, prices fall and the feedback loop is reinforced, steepening the former and lowering the latter. In Figure 6.1, the top X-axis reflects the degree of government action and incentives from high, at the outset of an infant technology/high-cost phase, to moderate, as prices fall and the diffusion of the S-curve steepens, to a maintenance phase, when a technology is widely diffused and government regulation oversees the new market with lowered incentives and a much lower technology price is seen. The bottom X-axis shows time. In 2021, some sectors are more advanced – such as wind power and solar – but are still to be widely adopted in all countries. Today, too many economic sectors are in the lower bottom left-hand quadrant and require much more effective policy action.

The left-hand Y axis shows the rate of diffusion from low to high. Government regulatory and economic policies should seek ways to speed diffusion, such as subsidizing EV charging stations, announcing phase-outs, regulating feed-in tariffs, and announcing high and rising efficiency standards for industry, construction and buildings, and processes, for instance.

In addition to the dynamics visualized in the matrix in Figure 6.1, there is a further important positive process that is often seen. As a technology becomes more widely adopted (diffused), prices fall, but the rate of internal innovation by leading firms, in efficiency, power output, scale, breadth of application, and so on, does not stay static or decline. Evidence shows, rather, sustained technology improvements and climate gains. This is clearly visible in the solar and wind space, with solar photovoltaic (PV) becoming progressively more efficient and cheaper, wind turbines larger, and generation less expensive. It is also visible in the ongoing improvements of battery technologies. We should not assume the rate of technological innovation is constant or must decline. Continuing innovation is possible.

Sustained dynamic innovation, price falls, and increased adoption rates can all support the net-zero goal. Pessimists might retort you have some gains in some sectors, but overall diffusion is a slow process. That was historically the case, but research on the rate of technology adoption has found it is accelerating, which is good news for the planet.

Diffusion comes in waves

Extensive work by Milner and Solstad (2018) on technology diffusion for 20 key technologies, drawing on 90,000 observations from 1820 to 2008, finds that technology diffusion occurs in waves. Moreover, they find that the gradient of slope has increased wave to wave, as per Figure 6.2. Assuming they are correct, and we are in a technology and now greening wave of unrivalled steepness, this gives cause for optimism that the crucial rapid adoption internationally of green technologies may be seen in lower-income as well as advanced economies.

Milner and Solstad's (2018) conclusion that 'competitive pressures in the international system thus generate critical incentives in the face of powerful domestic resistance to new technology. … systemic change may lead to waves of technology adoption in many countries' describes the dynamics of an ongoing rapid technology diffusion wave, one which started as an IT-computing-digitalization wave and is transitioning into a green technology wave, which will force countries to reach towards similar goals and adopt new technologies, thereby speeding the shift across the globe.

This description seems intuitively correct; we all see the rate of diffusion accelerating, not slowing.

Once the reality of climate change in a country, firm, or community is accepted, leaders with foresight who understand the need to mitigate climate change risks will grasp the economic opportunities of the new era and push their country, firm, or community to make the necessary changes in policy planning and to shift investments, set new priorities, and import technologies as needed so as to rush towards, not away, from the future. A virtuous planetary cycle of competitive political, economic, and technology diffusion pressures can occur.

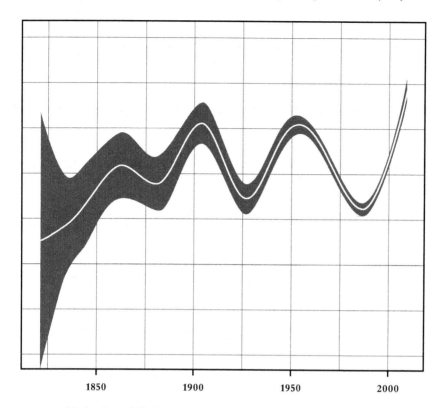

FIGURE 6.2 Technology diffusion as waves 1820–2006

Source: Milner and Skolstad, 2018.

Innovation is not episodic but, rather, is progressive and iterative

The history of innovation and diffusion is not a series of discrete eureka moments followed by struggles for adoption and implementation. The history of innovation has key junctures of discovery, but after this point there are multiple smaller breakthroughs and continuous innovation that speeds utility and efficiency. Inventors and innovators pursue what is called in IT the minimum viable product (MVP) approach, and this process itself drives diffusion rates, innovation, and progress.

This process of innovation is an engine of growth, rewarding innovative firms that iterate and improve with MVP processes, as teams seek again and again to gain a commercial edge. This green reindustrialization and retrofit is not a matter or waiting for the perfect answer or product. Rather, it is a matter of constantly innovating, adjusting, and reiterating (see Box 6.1). When critics say product A or process B is not good enough or is too costly or too inefficient for GHG impact X, remember that innovation and diffusion is not a static process but a continuous one that is iterative and dynamic and that can be accelerated.

BOX 6.1 PURSUING A MINIMUM VIABLE PRODUCT

Dynamic firms innovate using the MVP approach. An MVP approach is based on the premise that a new product can provide enough value to a customer to use but the customer then provides feedback on what to improve about it. This approach helps avoid creating products that customers do not want or need. For example, in an MVP approach, an inventor or firm has an idea for a new device, process, or product (e.g. a vehicle, a wind turbine, a solar design). The firm identifies an application or a niche or a need and designs a product, which is then released. The MVP version of the product has just enough features to be usable by early customers, who can then provide feedback for future product development. A focus on MVP development potentially avoids lengthy and unnecessary work and speeds innovation rather than slowing it. The next iterations of the product are more useful, efficient, productive, and profitable.

This is the history of technological innovation and diffusion. It is not a matter of a single breakthrough followed by an attempt to sell only that first early, expensive, limited product again and again. Steve Jobs created the iPod, a breakthrough only early adopters bought and few thought would be disruptive of the music industry. Yet through progressive iterations and the creation of related products (iTunes and so on), the invention and the MVP approach sped diffusion and adoption while at the same time making the products themselves progressively better, more desirable, and more versatile.

Most breakthroughs and innovation cycles mimic this rhythm. Some can get stalled or have difficulty in the early stages of diffusion (such as EV in agriculture). Others, with state intervention and support, move faster (such as wind and solar PV).

The role of government in supporting technology diffusion is to empower 'the green innovation machine … to activate private innovation forces for ecological transitions' (Veugelers, 2016).

Major, dramatic, and ongoing shifts are needed in all industries and sectors

Some policies aimed at making the transition to net zero across different sectors are working, while others have barely begun to work. Still others are scarcely altering current destructive, business-as usual-behaviours. These factors raise serious doubts about securing Paris Agreement goals.

Government policies must work to ensure alignment, to speed the rate of diffusion of net-zero technologies, systems, and industrial approaches. The following sections look at the progress being made and the many challenges in different industrial sectors undertaking the transition to net zero. Some sectors and markets, like electricity generation in Europe, and solar PV and wind in Europe, China, and certain

US states, are dynamic and evolving rapidly after years of governmental focus and shifts in incentives and regulations. Other industries and sectors, such as transport, are seeing progress, but more urgency is needed. Worryingly, other massive areas of our economies, such as agriculture and construction, lag terribly and could cause us to fail to reach global decarbonization goals.

An end to fossil fuel subsidies

Speeding up the rate of the global energy transition − away from fossil fuels to renewables − is essential if we are to stabilize global temperatures and avoid a dystopian hothouse future. Far more aggressive and immediate steps towards a phase-out of subsidies for fossil fuels must be taken. The subsidies distort markets and damage the planet. In 2015, fossil fuel subsidies amounted to US\$5.3 trillion (IMF, 2019) or 6.5 percent of global GDP, a staggering misallocation of resources with terrible consequences. If fuel prices had been set at appropriate levels in 2015, the IMF estimates global CO_2 emissions would have been 28 percent lower, fossil fuel air pollution deaths 46 percent lower, government revenues higher by 3.8 percent of global GDP, and net economic benefits (environmental benefits less economic costs) would have amounted to 1.7 percent of global GDP, or US\$1.62 trillion annually. All governments need to phase out these subsidies as a matter of urgency.

An energy transition underway

COP26 and governments must take additional steps to ensure 'current and new clean energy technologies are rapidly supplying all the growth in energy demand. Energy policies can reshape markets, business models and patterns of consumption leading to a peak in fossil fuel demand in the course of the 2020s' (WEF, 2019: 5). To achieve the 2050 net-zero goal, governments must support exponential rates of growth in renewable technologies and solutions, and a dynamic policy framework, with emerging countries leapfrogging technologies. Such a rapid scenario is possible and can be seen in BP's (2019) rapid transition scenario and models by BloombergNEF (2019), the IEA (2018), IRENA (2019), the IPCC (2018), and others.

There are positive dynamics underway. Technology disruption can be seen in the solar and wind sectors and also in the batteries sector and their application to renewable integration technologies. With prompt policy action the disruption can move onto transport and other areas. As I have stressed elsewhere, incrementalism has to be rejected. This is a war for planetary survival. As the World Economic Fourm (WEF) states, achieving a rapid transition:

> will require a major coordinated effort of policy, technology development and behavior from all sections of society to drive change across the whole of the economy on the timescale needed to achieve the goals of the Paris Agreement.
>
> *WEF, 2019: 9*

Governments must use all the levers they have to act on supply, i.e. to bring fossil fuel peak demand forward. Once the peak is visible, this will trigger further market shifts, accelerating the energy transition speed (Carbon Tracker Initiative, 2020). Indeed, perhaps the peak is already upon us. Covid-19 has slashed demand for fossil fuels, while renewables continue to expand; in some countries they provided all electricity generation for months in 2020.

We can see the transition-diffusion process playing out in Europe, in coal and fossil fuels, solar PV, wind, and the car sector. In each of these cases, incumbents are being disrupted and stock prices impacted. The new narratives and the routes ahead are visible.

Coal is (almost) down and out – in some markets

In the first half of 2019, coal demand in Europe experienced the sharpest decline ever recorded. Electricity generation from coal-fired power plants dropped by nearly 20 percent. This collapse 'can be attributed to the use of renewables such as solar and wind, as well as a recent increase in the use of gas-based power' (earth. com, 2019). It has also been driven by the higher prices in the EU ETS seen in 2019/2020. Europe has passed peak coal: coal consumption dropped by 79 percent in Ireland and 44 percent in Spain. In the UK, coal power generation dropped by 65 percent, and the country has committed to completely eliminating coal by 2025. Today, Europe's dirtiest coal plants are no longer profitable due to carbon and power prices and operating costs.

European and North American disinvestment from coal is accelerating; most leading European banks will no longer lend to coal projects, and there is a trend away from coal and coal stocks. Pension funds, for example, are disinvesting in such stocks, and the SNL Coal Index dropped 53.5 percent in 2019, while the rest of the markets rose, despite a supportive US administration and industry-wide efforts to avoid debt (IEEFA, 2020). It may soon be the end for American and European coal producers. China must be the next to follow this path.

Peak fossil fuel is upon us. The oil majors appear to be recognizing the reality that a significant proportion of their reserves will become stranded assets as this shift to renewables accelerates. In the first three quarters of 2020, US and European oil companies wrote down US$145 billion in assets – the largest amount ever – as 'oil companies also face longer-term uncertainty over future demand for their main products amid the rise of electric cars, the proliferation of renewable energy and growing concern about the lasting impact of climate change' (*Wall Street Journal*, 2020).

We can see the disruption in the rise of renewables. Non-fossil sources made up nearly one-third of the growth in energy supply in 2018, and the amount of energy they produce continues to grow rapidly. BP estimates that solar and wind made up 27 percent of the change in total energy supply in 2017. If current solar and wind growth rates of 15 to 20 percent per year are maintained, the Carbon Tracker

Initiative calculates that renewables will supply all incremental energy demand increases (not just electricity) by the early 2020s.

A sunnier and windier outlook

The cost of solar PV electricity generation has fallen dramatically since the early years of the technology. In the last ten years the cost of solar-generated electricity has fallen by over 70 percent (IRENA, 2020). Costs continued the downward trend in 2020. Solar generation in 2020 was competitive with even coal-fired power plants in China, or oil plants in the United Arab Emirates. And costs continue to fall.

In 2020, Italy, Germany, and Japan had the highest share of electricity produced from solar PV (PVPS, 2019: 86) at 9.2 percent, 8.4 percent, and 7.8 percent, respectively. Some countries lead in this energy transition and others follow. China is one of the leaders, even as they have a huge task to achieve their net-zero 2060 goal.

China continues to lead on low-cost solar and pushed down solar power costs 9 percent in 2020. Today, 'new-build solar in the country is now almost on par with the running cost of coal-fired power plants' (*PV Magazine*, 2020b). In 2020, total installed solar in China stood at 246 gigawatts (GW). Under the Paris Agreement scenario, that will need to rise to 2,803 GW by 2050 (China NREC, 2019: 178). Reaching such a figure is not impossible if we look at the rate of increase in installed capacity in China year on year: 28 GW in 2014, 43.2 GW in 2015, 77.4 GW in 2016, 126 GW in 2017, and 175 GW in 2018 (CEC, 2020). The challenge is huge when one considers China's 2020 net-zero-by-2060 target, but the country's ability to make abrupt capital investment shifts in years not decades indicates the rapid shift to solar and wind in China will continue, and, pressed by policy action, prices will fall further and the process will accelerate. Other states already benefit from China's action on solar; their huge PV production capacity has slashed prices globally. This cheap solar PV will also cause other countries to leapfrog technologies, using Chinese products to achieve their own energy transition (see Figure 6.3).

Solar electricity costs have been falling at over 15 percent a year since 2009, and solar innovation and improvement continues rapidly. Evidence suggests solar PV costs will continue to fall as the steepening diffusion curve accelerates interlinked processes.

Markets and investors see these trends and are adopting a green narrative, driving solar stocks up, as they anticipate the Biden administration's green industrial policy rollout of spending and regulatory programmes, as well as success at COP26. For instance, in the days before the US election, solar ETFs such as Invesco TAN soared 143 percent from its low right after the lockdown in March. Other solar firms soared as well, such as Enphase Energy at 317 percent year-to-date (YTD); SolarEdge at 221 percent YTD; SunRun at 417 percent YTD; and Nextra Energy, the US's largest solar utility, up 25 percent YTD (oilprice.com, 2020). Is this just irrational exuberance (Shiller, 2010)? Perhaps. But perhaps not if you understand the public policy shifts underlying it, are in it for the long run, and can see climate

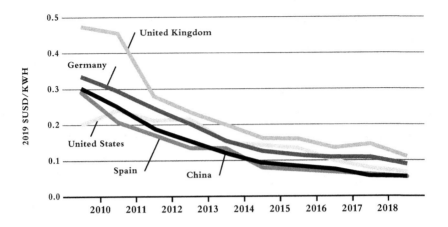

NOTE: All LCOE values are calculated based on project level data for total installed costs and capacity factors from the IRENA Renewable Cost Database, with other assumptions necessary for LCOE detailed in the source link below, notably an assumption of a weighted–average cost of capital of 7.5% real in the OECD and China and 10% elsewhere.

FIGURE 6.3 A sunny outlook for solar generation

Note: IRENA = International Renewable Energy Agency; LCOE = levelized cost of energy.

Source: IRENA (2020), Renewable Power Generation Costs in 2019, International Renewable Energy Agency, Abu Dhabi www.irena.org/publications/2020/Jun/Renewable-Power-Costs-in-2019.

crisis economics altering the market and future returns. The progress on solar is matched in the wind sector.

Wind is now competitive without subsidies

Wind power in 2020 has come a very long way from Dutch images of windmills grinding wheat in the eighteenth century. Today, wind turbine electricity is competitive without subsidies. The rate of diffusion has accelerated the transition, and market disruption is underway.

Scotland's progress, backed by Scottish Power (see Box 6.2), demonstrates what is possible. In the first half of 2019 alone, Scotland generated enough wind power to supply the entire country twice over. Wind turbines generated 9.8 million megawatts of electricity between January and June, enough to supply power to 4.47 million homes. Scotland only has 2.6 million homes (ScienceAlert.com, 2019). In 2020, Scotland was close to generating all its energy from renewables. During the same period, the UK managed the longest duration without relying on coal power since the Industrial Revolution of the nineteenth century.

The levelized cost of energy for onshore wind has fallen fast, as we also saw in solar PV and as shown in Figure 6.4.

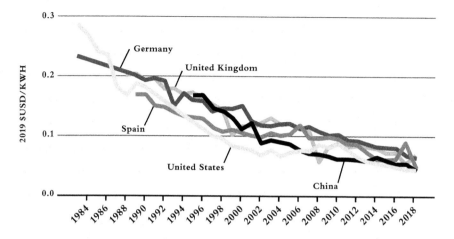

NOTE: All LCOE values are calculated based on project level data for total installed costs and capacity factors from the IRENA Renewable Cost Database, with other assumptions necessary for LCOE detailed in the source link below, notably an assumption of a weighted–average cost of capital of 7.5% real in the OECD and China and 10% elsewhere.

FIGURE 6.4 A profitable and windy outlook for turbines

Note: IRENA = International Renewable Energy Agency; LCOE = levelized cost of energy.

Source: IRENA (2020), Renewable Power Generation Costs in 2019, International Renewable Energy Agency, Abu Dhabi www.irena.org/publications/2020/Jun/Renewable-Power-Costs-in-2019.

BloombergNEF (2019) reports the lowest-cost projects were financed in Australia, China, Chile, and the United Arab Emirates in 2019/2020. In solar and wind, as key parts of the needed rapid energy transition, the future is being written today.

Again, we know policy matters for rate of transition. China, whose policy in support of wind generation is well established, leads adoption, with installed wind turbines producing 71.7 GW in 2020 (GWEC, 2021), which should rise rapidly to 366.4 a decade later (IRENA, 2020: 27). Projections suggest that under a 2050 scenario in line with the Paris Agreement, the total installed capacity must rise to 2,636 GW. This is not impossible given the rate of growth already seen in China (China NREC, 2019: 179), but there is good news elsewhere as well – even in America.

States in America's Bible Belt also show what can be done with good, depoliticized public policy. Perhaps surprisingly, Texas leads America in wind power, with 30.2 GW of capacity, the result of progressive policies implemented by George W. Bush when he was governor. I recommend driving across the northern Texas panhandle and up into Utah (see Figure 6.5). You will see hundreds of turbines. Kansas, another staunchly conservative state, generates fully 41 percent of its energy from wind, as does Iowa (AWEA, 2020).

FIGURE 6.5 Green Mountain Energy Windfarm, Texas

Source: Creative Commons, CC By-SA 3.0. See https://en.wikipedia.org/wiki/Brazos_Wind_Farm#/media/File:GreenMountainWindFarm_Fluvanna_2004.jpg.

First movers with the wind behind them

Increasingly, countries, markets, and first-mover firms are investing in wind and reaping the benefits. Firms such as Iberdrola, which owns Scottish Power (see Box 6.2), are pioneering wind power, and their stock performance reflects their leadership position.

BOX 6.2 WINDY GIANTS: IBERDROLA GROUP AND SCOTTISH POWER

Scottish Power, the utility that serves Scotland and which is owned by the Spanish Iberdrola Group, is advancing towards net zero in Europe by 2030 – an aggressive goal that places the firm at the forefront of the move from brown to green, and from green to greenest. The firm's emissions are a quarter of those of its European competitors.

In response to the 2015 Paris Agreement, and the SDGs for 2030, the firm embedded these goals into its corporate and business strategy. Aligning its firm, the company focuses its efforts on climate action (SDG 13) and on the supply of affordable and nonpolluting energy (SDG 7). The firm thus did just what external observers recommend: it aligned its business strategy with climate change goals and adjusted both its investment and risk strategies and its performance management processes (G30, 2018, 2020).

By 2019, the firm had already reached zero emissions in many of the countries including the UK, Germany, and Portugal (Iberdrola, 2021). To get to that leadership position, the firm invested more than US$100 billion in the last two decades in renewable energy, smart grids, and efficient storage. Has this investment paid off for this first mover? It appears so. The stock has significantly outperformed other utilities, and it reached an all-time high in 2020.

Wind ETFs also reflect the shift underway and have risen 30 percent YTD. Leading turbine manufacturers are on a tear: Vestas is up 200 percent in a year; Orsted is up 100 percent in a year; TPI Composites is up 250 percent in a year. The list is long, and the trend is increasingly clear. More and more investors are backing market and climate change narrative shifts with their money; they can see the future, and it is not one built on fossil fuels.

GE Renewable Energy is another company building the power infrastructure of tomorrow today (see Box 6.3).

BOX 6.3 NOT YOUR GRANDFATHER'S WINDMILL

In 2020, GE Renewable Energy introduced the Haliade-X12 wind turbine (Figure 6.6). At 260 metres (853 feet) high, with a 220-metre rotor, a 107-metre blade, and a 12-million-GW capacity constructed of composites, it is the most powerful offshore wind turbine in the world. The blades bend and flex

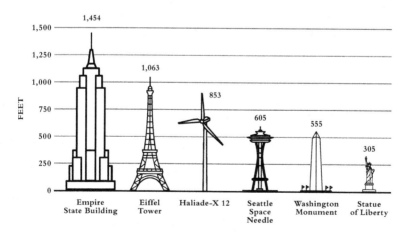

FIGURE 6.6 Towering turbines: The new green utility giants

Source: GE.

under the pressure of operation, with a capacity factor of 63 percent: this is the average power generated, divided by the rated peak power.

With massive blades, this turbine and other such giants can generate more power at low wind speeds. Its 67 GW hours (GWh) of gross annual energy production provide enough clean energy to power 16,000 European households and save up to 42,000 metric tons of CO_2, equivalent to the emissions generated by 9,000 vehicles in a year. GE Renewable Energy is among the leading turbine builders in the world, supplying offshore wind farms in the US and in Europe.

The story of this energy transition is only partial, but it is a positive one about effective governmental innovation policies, subsidies, support, and taxation spurring early-stage breakthroughs, early adoption, and investment. The diffusion S-curve is steepening. Prices are falling rapidly. Innovation is still happening with the feedback loops and innovation dynamics we require to achieve net-zero goals by 2050. It appears that:

> The era of carbon-intensive energy derived from the burning of fossil fuels is coming to an end, and a cleaner, more reliable energy future based on renewables like wind and solar will be the new normal.
>
> *Kortenhorst, 2019*

The energy transition is driving job growth. America's two fastest-growing occupations through 2026 will be solar installer (105 percent growth) and wind technician (96 percent growth) (*Forbes*, 2019). Solar and wind market growth, adoption, and maturation show what is possible when governments change the incentives structures, subsidies, and early adoption support. There are other positive examples of accelerating diffusion, such as the transport sector, which is undergoing disruption, redesign, and revolution.

An EV transport revolution in motion

Securing the transport transition requires a revolutionary shift away from petrol combustion engines to EVs. Cars alone contribute 15 percent of GHG emissions, and the transition is urgent. As the S-curve is steepened, country to country, EV prices will fall and innovation rates can be sustained. This process can be seen in Europe, with governmental action spurring market reactions, adjustments, investments, and individual behavioural shifts, all brought forward by pulling the diffusion curve towards us. In 2020, credible transition scenarios predict a rapid increase in EV sales from 2016, with a steep S-curve of adoption pushing market share as high as 90 percent by 2035 in many markets, but by no means all. Some markets, such as those in East Asia (with its hundreds of millions of motorcycles and cars) and the US (with its love affair with trucks and mammoth SUVs) lag and must be made to shift to gearless, faster EVs sooner.

Globally, the EV transition is at an early stage but is taking off, with sales up 40 percent globally year on year in 2018 and 2019 (IEA, 2020) and with Europe leading the way. EV sales accounted for 6.8 percent of Europe's total vehicle sales in the first quarter of 2020, up from a 2.5 percent share a year earlier, indicating robust rising demand despite the pandemic. There are numerous signals that the market will leap forward in the 2020s. The number of EV models available in the European market is set to reach 214 by 2021, more than doubling in two years from 98 at the end of 2019, according to the European Battery Alliance. With all these new models to sell during an era of declining overall vehicle sales, EU countries and the UK are moving to encourage uptake of EVs. Governments are setting phase-out timelines, ranging from 2025 to 2030 and beyond, creating structures of incentives and penalties. For instance:

- In Norway, the announcement of the most aggressive phase-out goal globally of 2025 electrified the market transition. With 15 incentive schemes to spur the shift, already in 2020 over 50 percent of cars in Oslo and 44 percent of the market overall are EVs. Charging stations are in place on almost every street in the city and are being put in place every 50 miles in the rest of the country.
- In the UK (phase-out by 2035), once the ban comes into effect, people will only be able to buy electric or hydrogen cars and vans. EV buyers can receive subsidies of 35 percent of the cost of an electric car (up to £3,500), 20 percent of the cost of an electric motorcycle or moped (up to £1,500), and 20 percent of the cost of an electric van or truck (up to £8,000).
- In France (phase-out in 2040), EVs are exempt from the GHG tax, there are subsidies of up to €7,000 for households buying EVs below €45,000, and there is a scrappage scheme of up to €5,000 for households and €2,500 for individuals.
- In Italy, EVs are tax exempt for five years from registration and get a 75 percent reduction after that period. Italy also has a bonus-malus scheme, where vehicles are subsidized up to €6,000 per car emitting less than 70 gigatons of carbon dioxide per square kilometre but are penalized by €2,500 per car if they emit above 250 gigatons of CO_2 per square kilometre.
- In the EU's largest market, Germany, several policies that were rolled out in 2020 seek to increase EV demand, including cutting the value-added tax on EVs by a third, giving new EV owners a ten-year tax exemption, and providing a €9,000 subsidy on EVs with a sales price under €40,000.

These programmes, and many others, exhibit the type of forward thinking and mix of incentives, penalties, and tax treatments needed to prompt behavioural shifts. It is European governments that are predominantly leading the way on the transport transition.

Other major markets, such as California, China (which has the world's largest number of EVs – 3.6 million in 2019), and Japan, are pushing forward with target phase-outs and related subsidies and policies. For example, polling in China shows how early adoption requires government policy support, with 32 percent of buyers saying EV purchases were being driven by government subsidies and benefits, 31 percent citing the driving experience, 17 percent citing total cost of ownership impacted by government support, and only 15 percent citing environmental factors (McKinsey, 2020: 12). Credible, predictable, ongoing policy adjustments hasten adoption and speed market dynamics, sending signals to manufacturers. Other countries must follow and apply this type of suite of policies to spur the transition. President Biden should look to Europe and apply what we know works.

The EV transition is running in parallel with the battery technology revolution.

Battery costs are falling fast, speeding the transition and steepening the S-curve

Battery costs are falling fast as this technology – and new variants and continuing innovations and research – continue to increase battery capacity and efficiency and push down prices. Battery pack prices have been falling swiftly from US$236 per kilowatt-hour in 2017 and are projected to get to at least US$110 per kilowatt-hour by 2025 (see Figure 6.7). They will soon get to below the US$100 number thought crucial to speed diffusion. The latter figure is considered the jumping-off point for much wider adoption of the technology. We are almost there.

Falling battery costs help accelerate EV transition, and as the transition continues, demand for EV batteries is forecast to accelerate very rapidly indeed. Demand for EV batteries stood at 110 GWh in 2020, and will rise to 1,910 GWh by 2030, leaping to 5,910 GWh by 2040 and to 6,530 GWh by 2050 (*Faraday Insights,*

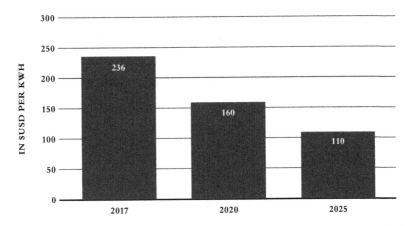

FIGURE 6.7 Time to charge up?

Source: McKinsey, 2020; Greentech Media, © Statistica 2020.

2020: 2). The national security and potential economic value of rare earth metals and the companies that mine them in China, the US, Myanmar, and Australia reflect this market shift. Battery manufacturers are reaping the benefits, too, as the market will be worth US$85 billion in 2025 and continue to rise from there. Household names such as Tesla (up over 300 percent in one year), BYD (up 200 percent in one year), and Cytokinetics (up 300 percent in a year) as of November 2020 are reflecting the bullish outlook for EVs and battery tech.

This rapid take-up of EVs will reinforce the energy transition. Ultimately, the EV fleet will become part of distributed EV-to-grid (EV2G) systems, further accelerating the change in technologies, generation, and storage systems (see Box 6.4), saving money for utilities and for consumers.

BOX 6.4 EVS, SOLAR PV, AND DISTRIBUTED STORAGE AND PEAK SHAVING

A major challenge of renewable energy is storage and demand – i.e. how to store enough electricity to ensure supply meets demand during peak periods as solar PV and wind increase their share of supply. Both renewables are episodic in output – i.e. they are subject to the sun shining and the wind blowing – and those periods do not reliably line up with peak demand periods. So, what to do? Link EVs, residential solar, and battery storage. There will be a move away from centralized generation, revolving around a small number of power stations, and towards a decentralized form with cities, towns, and houses acting as their own mini power stations, with solar PV and battery storage. 'EVs, residential solar and storage offer an efficient and cost-effective way of achieving this type of energy system' (*PV Magazine*, 2020a).

EV2G allows electric and hybrid cars to return stored energy to the power grid. This allows utilities to respond when demand for electricity is highest and many EVs are parked while their owners are having their breakfast or dinner. As the number of EVs increases rapidly across the globe, EV2G will be an important facet of the renewable power solution. 'EV2G … will bring advantages both for the EV user and for the grid' (Castellanos *et al.*, 2019). Decision Analytics estimates that in Texas, one of the leading US states in renewables generation, 1,000 megawatts of distributed storage would save the state US$344 million a year by cutting the cost of distribution and transmission. The combination of wind, solar, and batteries (including EV batteries) cuts costs by cutting peak demand on the system (*PV Magazine*, 2020a). Texas lawmakers understand the link between renewables and cost and have passed bills through the state legislature to support it. Other states and localities should follow the lead of Texas and plan accordingly.

A similar EV revolution is required, and has only just begun, in road freight transport. Major manufacturers have announced EVs designed for the 80 percent of freight transport that is delivered over less than a 250-mile radius. The revolution is in its very early stages, but if the trajectory follows that seen in auto EVs, there is reason for cautious optimism. Here, too, governments must structure their responses to hasten the shift, steepen the diffusion adoption S-curve, and pull it towards us.

In summary, the energy revolution is underway. Unfortunately, such urgency is not seen in other massive sectors of our economies, the worst example of which is perhaps agriculture, and failure here threatens to defeat us all.

Agriculture and farmers find the transition tough

GHG emissions from agriculture and land use are a major contributor to global warming. During 2007–2016, agriculture, forestry, and other land use accounted for 13 percent of CO_2, 44 percent of methane (CH_4, with 28 times the warming effect of CO_2), and 82 percent of nitrous oxide (N_2O, with 80 times the warming effect of CO_2) emissions from human activities globally. Agriculture and land use together produce 23 percent of GHG emissions. Including all food production raises this figure to between 21 and 37 percent (IPCC, 2019). Because of this magnitude, speeding the net-zero transition in land use and agriculture is essential and must be a core component in the transition story and national green industrial processes and planning.

Adaptation and technology improvement

There are pathways to a better future for our farmland, animals, and diets. These pathways should include, in descending order of GHG impact and importance:

- Requiring zero-emission farm machinery and equipment (537 metric tons of CO_2 equivalent [$MTCO_2e$])
- Better selection and breeding of new crops (508 $MTCO_2e$)
- Shifting practices in rice fertilization (449 $MTCO_2e$)
- Changing animal healthcare (411 $MTCO_2e$)
- Altering animal feed mix (370 $MTCO_2e$).

These are just the top-five targets of many steps that could cut emissions by almost 2.5 billion $MTCO_2e$ by 2050 (McKinsey, 2020). All these adaptations have varying costs. Today, unfortunately, farming practices in most countries lag far behind what is needed to achieve necessary GHG savings. Markets will not get it done alone. Most of the required measures are unlikely to be driven by market forces. Policy interventions are likely to be needed.

Scrap that old tractor

Take, for instance, improving farm machinery efficiency. Most farm machinery remains inefficient and more polluting than current consumer vehicles. Here, as in so many other fields, spurring market shifts and farmer response requires enforced government mandates for efficiency and scrappage. Governments must set a time-table for the phase-out of combustion engine use in agriculture, just as many are doing with internal combustion automobiles. Only governments can push farmers to stop using that old, rusting, inefficient tractor and replace it with a new, low emissions or EV. The EVs are already being built by innovative manufacturers. Governments need to set timelines that speed their adoption (see Box 6.5).

BOX 6.5 EV TRACTORS' DEEP FURROW

California-based Soletrac designs and sells electric tractors that can replace the diesel vehicles. Compared to diesel tractors, EV tractors are cleaner, quieter, more efficient, and less expensive to use. Importantly, they provide maximum torque and maximum power, regardless of speed. Operating costs will be lower than diesel (certainly when carbon is priced properly). As with other EVs, tractors have just a single moving part in the motor, whereas diesel tractors can have more than 300 parts. Having a quiet and more efficient tractor, espe-cially when used for several hours per day, will be a compelling argument for modern farmers to make the transition. The EV tractor market is small at pre-sent but is set to grow rapidly. Governments should speed the transition with incentives and medium- and long-term phase-out announcements. That will jump-start the farm transition.

Major manufacturers are moving forward alongside Soletrac. Kubota of Japan already sells a series of 24- to 47-horsepower electric tractors. Fendt of Germany produces a 70-horsepower model running five hours on the battery and getting up to an 80 percent recharge in four hours. Lindt of Austria has an EV model in trials. Pilot models of electric tractors are also under development by John Deere, Case/IHC, and several other European makers. The industry is gearing up for the shift. As the John Deere company states, 'We believe in elec-tric tractors. 100%' (Future Farming, 2020). Governments need to accelerate the timeline for transition.

Ultimately, it is expected that many farmers will use solar PV and wind to recharge batteries, cutting costs and GHGs for operations still further.

New crops to lower emissions

Consider the genetic challenge for farming. Given too many of us continue to eat too much meat, genetic selection needs to be harnessed to produce feedstuffs that generate less methane when consumed by cows and sheep. This is not impossible;

experiments have shown that selective breeding in single herds can reduce methane by as much as 20 percent (McKinsey, 2020; Pickering *et al.*, 2015). In countries with huge ruminant populations – such as Australia, where 17 percent of the country's GHG emissions come from animals (most of which are sheep and cows) – genetic selection for lower methane could have a rapid and significant impact (Cole *et al.*, 1997). Here, too, governments must lead, educate, and provide R&D and other support to speed up changes in breeding and animal methane emissions.

Growing rice but limiting methane

Take rice as an example. More people rely on rice for subsistence than any other crop. Unfortunately, rice paddies are perfect places for methane and nitrous oxide (N_2O) production as they are warm, muddy masses that bubble with decomposition and bacteria. Yet we know that farmers who adopt better fertilization practices can reduce methane and nitrous oxide produced in rice farming (Linquist *et al.*, 2012). Chinese farmers are already altering their practices by draining paddy fields in the middle of the season, changing cultivation methods, and cutting methane production compared to the 1980s (Qin *et al.*, 2015), yet rising climate temperatures threaten to increase methane from rice by as much as 60 percent in the future. Here, education and research matters, not just, or mainly, mandates.

Better farming practices mean better outcomes

Reforming farm practices is urgent. GHG reductions could be achieved if farmers managed fertilizer use better by avoiding overuse, runoff, and nitrous oxide production. As with many examples, we know what works – cover crop planting, no-till techniques, silvopasture (the practice of integrating trees, forage, and the grazing of domesticated animals in a mutually beneficial way), and alley cropping. All these techniques produce environmental GHG dividends. Other simple practices, such as riparian barriers and windbreaks, also affect GHG emissions. New techniques and approaches are also being developed. For instance, as the carbon price rises, new rocky, dusty, partial solutions may become economic and may impact GHG emissions (see Box 6.6).

BOX 6.6 A ROCKY, CARBONATED GHG SOLUTION

Research published in *Nature* (Beerling *et al.*, 2020) demonstrates that spreading rock dust on farmland could remove huge amounts of GHG from the atmosphere. The chemical reactions in the soil degrading the rock turns it into carbonates within months, locking CO_2 into the soil. Treating half the world's cropland with rock dust would lock in 2 billion tons of GHG each year, equivalent to the combined emissions of Germany and Japan. Many countries already have huge stocks of unused, already mined basalt rock, notably the US, China, and India, which are major GHG polluters. The cost ranges from US$80 in India to US$160 in the US per ton of CO_2.

When carbon prices rise – as they should – the calculus can shift dramatically in favour of this type of farming practice. This dusty, innovative approach would pay real climate dividends to farmers and to the planet. Such practices 'can scale up and are compatible with existing land uses', according to Beerling, the lead author on the study.

Existing mining activities in many parts of the world already produce basalt dust as a waste. That waste may soon be worth a lot and help sustain mining industries through direct payments for this dusty, abundant (basalt is among the most common rock on earth) GHG offset material.

Farmers know many of the production and technological answers. Yet, land use and agricultural incentives are still not aligned with climate crisis goals. Neither Europe nor the US fully aligns its support of farmers to incentivize alteration of poor practices and adoption of new and existing methods that deliver GHG reductions. Europe has begun to reorient payments in this manner. My farming brother-in-law is paid to plant hedgerows, to leave land fallow, to plant native trees. European farmers are paid to maintain permanent grassland and to dedicate 5 percent of farmland to trees, hedges, and fallow land. These changes have resulted in a 2 percent per year reduction in GHG emissions (EU Commission, 2021). Farmers who fail to apply these standards receive lower incentive payments and local penalties. These structured incentive links should be extended and strengthened across all agricultural support to increase the pressure on Europe's farmers to change their practices.

US agricultural support is badly structured, much less directive, and less effective compared with Europe. In 2020, US conservation support and payments were separate 'feel good' programmes, not embedded within farming payments and the incentives structure; today most American farmers stick with counterproductive practices. You can see this when you drive across America, where scale is everything. There are no hedges, no trees. The bigger the field, the better. Monocultures of genetically modified crops, fed with fertilizer and made pesticide resistant, predominate. Better agricultural and land use practices would pay swift dividends. As the Institute for European Environmental Policy has noted:

> Agriculture along with other rural land-using sectors, is unique in its ability to both reduce its own emissions, but also increase carbon removals from the atmosphere, and contribute to emission reductions in other sectors through the substitution of carbon intensive materials and energy.
>
> *IEEP, 2019: 7*

BOX 6.7 FRANCE'S SHADY, DAPPLED, WOODED SUCCESS STORY

In 2019, a remarkable one-third of France was covered in forest. While much of the world is losing its wooded landscape, France has decisively moved in the other direction, creating a vast and growing carbon sink. Supported by public policy and government incentives, there has been a concentrated reforestation effort, coupled to a decline in farming. In the last 30 years, France's forested areas increased by 7 percent.

The French success story has been part of a collective effort of private individuals and public forestry initiatives working together. Since most forests are on private land, landowner buy-in was essential and has been achieved. In 2019, French forests covered 16.4 million hectares, and the area is growing year on year. French forests are reclaiming old agricultural and industrial wastelands to fuel their growth.

France is setting aside and protecting and growing its forests. The newest addition is the Baronnies Provençales, in Provence. Established in 2015 and covering more than 1,800 square kilometres (approximately 700 square miles), this mixed forest of oak, pine, and beech is a testament to France's dedication to regrowing its ancient forests. Other countries within and outside Europe should learn from the French wooded success story.

Landowners and farmers need to be incentivized to change their practices. Large steps forward are in fact possible, as the French reforesting example shows (Box 6.7).

Align incentives with environmental goals

States should link all future farmer support payments to ongoing and required progressive changes in farming practices. Making the shift towards net zero requires that environmental and planetary stewardship be not an add-on or afterthought in support of farmers but, rather, an integrated part of land-use and farming practice. Farmers, and we as consumers, should be considered and should pay the real planetary price of that burger, lamb kebab, and pork taco, and be part of the process of internalizing the cost of carbon.

We need to change what we eat

At present, approximately 50 percent of the globe's usable land is dedicated to agriculture, and about 30 percent of cropland is used to grow grain for animal feed. As the demand for meat increases, so does the speed of deforestation and the removal of carbon sinks, thereby increasing GHG emissions and adding to the pressure on our climate and agricultural production, which suffer as a result. Individuals must consider how much and what we eat if we are to have a healthier, more sustainable tomorrow.

We need to alter our choices and our diets if we are to avoid adding 593 million hectares – twice the size of India – to feed the world's expected population of almost 10 billion in 2050. The world's population is not just growing, it is eating more – approximately 8 to 12 percent more – each year. For instance, the consumption of meat in Spain increased between 1970 and 2005, with average annual meat consumption per capita rising from 11.7 kilograms to 65 kilograms (Rios-Nunez and Coq-Huelva, 2014). Globally, if current consumption patterns continue, total meat consumption will increase by 72 percent between 2000 and 2030. This is terrible news for our planet and at direct odds with GHG goals.

Therefore, we need to eat less meat (especially bovine meat) if we are to avoid further degrading the land and instead reverse the trend and cut GHG emissions in the next three decades. If we ate less meat, we 'could reduce global mortality by 6 to 10 percent and food-related GHG emissions by 29 to 70 percent compared with a reference scenario in 2050' (Springmann et al., 2016: 4146). Getting there is at once easy (pick the salad, not the beefburger), and yet very difficult (most of us pick the burger). As Springmann et al. (2016: 4147) note, 'less than half of all regions meet, or are projected to meet, dietary recommendations for the consumption of fruit, vegetables, and red meat, and exceed the optimal total energy intake'.

Many more of us need to go on a diet and change what we eat, while all farmers and producers must shift practices and strategies to achieve net zero and align their businesses to the local achievement of the goal. At present, the average global consumption of ruminant meat (sheep and cows) is three times the recommended daily amount. Americans will fight for the right to eat their hamburgers. In the end consumers will need to pay slightly more for their burgers, and much else besides, and be incentivized to eat less, more mindfully, and feel better for it. We need to stop. Can we?

There are some small positive signs visible in our eating behaviours among a small but growing percentage of us. The leap in demand for meat substitutes appears to be one such indicator of a shift just beginning (see Box 6.8).

BOX 6.8 BEYOND MEAT'S MUSCULAR OFFERING, OTHERS JOIN THE GRILL-OFF

Beyond Meat is one of a growing number of meat substitute companies, and supplies its burgers and meatless products to groceries, restaurants, and fast-food joints. The firm's share price is up 60 percent in 2021, despite the pandemic's crimp on restaurant sales. Despite the headwind, retail sales of its products continue to grow at 40 percent per annum. Its stock reflects its leading-edge role, trading at exuberant levels of 738 times forward earnings, which may be hard to sustain in the medium to long term.

Major firms are rushing to fill the growing demand. Consumer goods behemoth Nestlé launched a rival to Beyond Meat's burger patty using similar pea-protein technology. Upon further examination, formidable new competitors such as Tyson Foods, Unilever, and many others have debuted plant-based burgers since Beyond Meat went public.

Beyond Meat will certainly not have this promising niche to itself. But that is less important than the trendlines and the shift that is gradually underway. The demand for meat substitutes is clearly growing among consumers, as reflected in the offerings of major fast-food operators, from McDonalds to Greggs in the UK, both of which have added meatless offerings to their menus. In addition, increasingly there is a growing expectation among consumers that the major firms that buy meat should act as better climate stewards. In 2020, in the UK, for instance, 21 major firms demanded action on legal and illegal deforestation (BBC, 2020c) through mandated comprehensive reporting down through the supply chain.

Shifting technologies and business practices also support GHG goals and can help ensure our diets are less planetarily destructive. Farsighted firms such as Brewdog and Walkers are taking the lead and closing the loop in their food manufacturing processes and GHG emissions (Box 6.9).

BOX 6.9 BREWDOG'S SODDEN SOLUTION AND WALKERS' CIRCULAR, CRUNCHY, SALTY LOGIC

Brewdog, the fourth-fastest-growing food and drink company in the UK, with sales up 114 percent in 2019, is taking the lead on climate change with its business strategy to 'Make Earth Green Again', a plan to plant 1 million trees in Scotland and make the firm double carbon negative. That is, until the trees are grown and removing GHG emissions from the air, the firm will purchase offsets to remove twice the firm's annual emissions. The firm's purchase of over 2,000 acres of highland moor will be split between 1,500 acres of native woodland and 500 acres of peatland restoration. Peatbogs are a major carbon sink. Sustaining them locks away carbon.

Leadership always matters. The firm's strategy is championed by Mike Berners-Lee, its chief executive, who shifted the firm's strategy after hearing David Attenborough talk of the climate change crisis. Brewdog is demonstrating that you can align your business with climate goals and be highly successful commercially. This is not about corporate social responsibility (although I am all for that); this is about business practices and strategies for the long haul that deliver for consumers, employees, and shareholders.

Walkers, manufacturer of the eponymous and ubiquitous crisps in the UK, are moving to a circular manufacturing process, one that will cut GHG emission by 70 percent. The firm will begin using CO_2 from beer production and mix it with potato waste from crisp manufacturing, turning it into fertilizer. This will then be spread on fields to feed the next year's potato crop for crisp production. The approach is innovative and was supported by government R&D grants (green industry in action) in its early stages. The firm already uses anaerobic processes to digest potato waste and generate methane, which is used to make electricity for crisp frying. This new process will use the leftovers from the digesters to create the fertilizer.

Walkers is practising carbon capture and will move the firm towards becoming carbon negative. The firm, which is owned by PepsiCo, is a UK pioneer in the practice of the circular economy, in which wastes are turned into raw materials while positively impacting GHG emissions. Walkers' business approach, mindset, and climate change stance are to be commended and should be copied where possible (BBC, 2020a).

Protestors against plastic waste complain they have yet to see PepsiCo take a similarly progressive companywide approach to GHG emissions reductions. In 2020, PepsiCo announced a target to source 100 percent renewable electricity across all its company-owned and controlled operations globally by 2030 and across its entire franchise and third-party operations by 2040. If achieved, this could cut 26 million metric tons of GHG from the firm's footprint (PepsiCo, 2021).

It is imperative to get increasing numbers of firms, farmers, and consumers to shift actions and adjust their narratives, climate stories, and understandings. Advocates for action need to use facts and dialogue to answer sceptics' views about climate change. Research shows it is the climate change doubters who are most resistant to changing their diets. As I have observed, if you want to change how people act, they need to first understand and internalize the facts. They then might begin to consider the dangers of doing nothing on climate change, before being encouraged to reflect on their personal choices (Boer et al., 2013). Do not browbeat the meat lover, the beer drinker, the crisp eater. Talk about the facts and challenges of climate change, and then, once the facts are understood and agreed, discussion on personal choices will have a greater chance of changing minds. We know this from our own discussions across the family dining table. Lecturing and lambasting is not the best route to changing minds and habits.

Industrial manufacturing and processes

Getting to net zero by 2050 is especially difficult in industrial manufacturing and processes. In a wide array of industries and activities, continued mitigation efforts are urgent and essential, and thus far too limited in application and effect. As the 2018 IPCC report notes: 'An absolute reduction in emissions from industry will require deployment of a broad set of mitigation options beyond energy efficiency measures' (IPCC, 2018: 743).

Efficiency gains can be secured, as can emissions reductions, through strategies such as fuel and feedstock switching, carbon capture and storage (CCS), material use efficiency, new product design, recycling and reuse of materials and products, and innumerable other steps.

But as of 2021, industry-related GHG emissions continue to increase, even as the total share of manufacturing in the global economy has declined. Manufacturing has fallen as a percentage of the economy. The sector is producing more with fewer, more productive workers but is also increasing GHG emissions. In 2018, the emissions from industry were larger than the emissions from either the building or transport sectors and represent just over 30 percent of global GHG emissions. Regionally, Asia is the biggest culprit. The IPCC has underscored that:

> The energy intensity of the sector could be reduced by approximately up to 25% compared to current level through wide-scale upgrading, replacement and deployment of best available technologies, particularly in countries where these are not in practice.
>
> *IPCC, 2018: 243*

Innovation is expected to result in a further 20 percent cut in energy intensity.

Looking longer term, step changes should include a shift to low-carbon electricity, radical product innovations (such as alternatives to cement), or widespread use of CCS. The IPCC makes clear that 'There is no single policy that can address the full range of mitigation measures available for industry and overcome associated barriers' (IPCC, 2018: 744). Cement production shows the climate change challenge of the task ahead.

Taking concrete steps proves difficult

Concrete is the most widely used building material in the world. The production of cement (a key component in concrete) is a significant contributor to climate change. Each year, over 4 billion tons of cement are produced, which produces approximately 8 percent of GHG emissions. Little progress has been made in cutting these GHG emissions. A rising carbon price would begin to shift incentives towards mitigation and carbon capture technologies, some of which do exist (Cordis, 2020) but most of which have not yet been implemented. At present, the new technologies remain niche options and experimental. This must rapidly change.

Construction business practices are particularly resistant to change. The building industry's profitability generally depends on fast construction schedules that do not internalize life-cycle environmental and carbon costs. The industry is localized and weakly regulated in many regions and countries. Ultimately, governments must force the industry to reform and redesign its processes, supported by regulation to reduce or capture GHG emissions and apply more circular industrial processes that internalize GHG costs. Yet alarmingly, in 2020, two-thirds of countries have no building efficiency standards at all (UN GlobalABC.org, 2021), and most new construction will take place in those countries.

More effective governmental regulation and oversight are essential. Strict regulation and enforcement of increasingly stringent building codes for commercial and residential building are required if GHG reductions are to be achieved via construction and in the efficiency of buildings during their lifespans. Greening of the entire construction industry will have to be accompanied by the widespread use of green concrete as the material of choice for general construction. It is not that nascent solutions are absent. They do exist. It is rather that they are not yet widely adopted. The use of fly ash in the mix, for instance, can avoid, kilo-for-kilo GHG emission that would otherwise result. Additions like recycled plastic also cut GHG emissions and strengthen concrete as much as steel does (*The Guardian*, 2016). Green cement technologies are being developed. Leading firms have made commitments to reform, but they are still a distinct minority (see Box 6.10).

To get to net zero, industrial firms including cement manufacturers will need to use carbon capture, utilization, and storage (see Box 6.11). They will also have to commit to using circular economy models that transform waste into fuel. The real test is in the firm-by-firm implementation of stated goals and adherence to TCFD requirements, prompted by markets that should reward compliant firms for action now and punish those that fail to act.

BOX 6.10 CEMEX'S NET-ZERO GOALS

In 2020, Cemex, one of the world's largest producers of cement, announced new climate and net-zero targets. The company committed to a 35 percent reduction of CO_2 emissions by 2030, in line with the goals of the Paris Agreement. In 2020, the firm also established the goal of delivering net-zero concrete by 2050. As Cemex CEO Fernando Gonzalez stated:

> In our business, we believe concrete – our end product – has a key role to play in the transition to a carbon-neutral economy. … This is why we have defined a more ambitious strategy to reduce CO_2 emissions by 2030 and to deliver net-zero CO_2 concrete by 2050.
>
> Cemex, 2020

The firm has laid out a CO_2 roadmap that includes investing in energy efficiency, using alternative fuels, expanding the use of renewable energy, and increasing the use of new material in cement.

BOX 6.11 CARBON CAPTURE AND STORAGE

The economic viability of CCS technologies rests fundamentally on the price of carbon and the price paid for the offsets by purchasers. With carbon priced at between US$1 to US$7 per metric ton in the US, or US$20 per metric ton in the EU, these projects are expensive tests of concept. When carbon prices rise to US$100 per ton and up, the calculus changes dramatically in favour of a rapid build-up of CCS projects.

For instance, Chevron's demonstration project, the Gorgon Carbon Dioxide Injection Project – is one of the world's largest integrated CCS projects in operation. Once fully operational, the project is expected to reduce GHG emissions by nearly 5 million metric tons per year, approximately equivalent to GHG emissions from annual electricity usage in 620,000 US homes. Clearly, a rapid rise in the carbon price will incentivize Chevron and others to rapidly build such projects, once the long-term rising cost of carbon is thought regulatorily certain and net-zero pledges are being translated swiftly into measurable, enforced glidepaths to net zero.

Investors and major companies across the world are poised and ready to commit resources to CCS once pricing incentives are altered by governments and regulators. Many of the leading-edge firms are not oil majors but have access to huge amounts of capital, know-how, and the ability to execute once the policy landscape and investing horizon is clearer.

Critics retort that CCS facilities are infant technologies not yet rolled out in large enough numbers to impact the climate outcome. This is true, but the investor community is poised. With the right incentives and pricing signals, the shift and a rapid buildout will be seen. As with other new technologies, the time from invention to widespread diffusion and adoption is falling. This will be the case for CCS if states set the right price signals and begin a fundamental shift and more rapid decarbonization. Ideally, each government must set goals for carbon pricing and support the rapid implementation of CCS as part of their national plans to achieve net zero by 2050.

Every GHG-polluting country needs to have clear goals and prepare for the construction of CCS and their operation as part of a transition plan. The UK Climate Change Commission estimates that annual capture and storage of 75 to 175 metric tons of CO_2 ($MTCO_2$) will be required in the UK by 2050, which would require a major CO_2 transport and storage infrastructure servicing at least five clusters and with some CO_2 transported by ships or heavy goods vehicles (CCC, 2019). Here again, we know what is necessary and we have the technology. Fix the price incentives and regulatory guardrails. Markets will do the rest.

Similar GHG emissions reduction challenges face industrial production in many other core sectors, such as iron and steel, which produced 3.7 billion $MTCO_2e$ in 2019, and plastics and rubber, which produced 1.4 billion $MTCO_2e$ in 2019. The production of wood, aluminium, and other metals added to GHG emissions. All such sectors need to modernize and redesign processes, increase efficiency, and cut emissions to close the loop and capture emissions, while carbon pricing signals must rise and alter the calculus and favour modernization and the rapid closure of old plants and adoption of new approaches.

If the industrial processes that built our modern cityscape are hard to change, so too are our habits of travel and all that comes with it. As much as fully 10 percent of global GDP comes from tourism, and a considerable proportion of that tourism and business-related travel relies on our addiction to flying. This, too, must change.

Flying must reflect the cost of convenience

Aviation is responsible for over 2 percent of all GHG emissions. While, pre-Covid-19, 4.5 billion passengers flew in 2019, this is expected to double by 2037. If aviation were a polluting country, it would be one of the largest in the world, ranked between Japan and Germany.

Flying today does not reflect the cost borne by the planet. Currently, airlines are bound by the Carbon Offsetting and Reduction Scheme for International Aviation (CORSIA), a market-based mechanism coordinated and overseen by the International Civil Aviation Organization. The agreement covers 77 percent of international aviation and could result in offsetting of 2.6 billion $MTCO_2$ through 2035. The aspirational goal is to make most of the growth in international flights carbon neutral. The agreement applies from 2021 and ramps up in three stages, requiring increases in airline efficiency, the use of biofuels, and the purchase of offsetting, according to a common methodology and metrics, by carriers flying between CORSIA states. The agreement will be reviewed every three years. It is quite possible that the effects of Covid-19 will both cut emissions in 2021 and require a review of CORSIA.

We cannot yet judge CORSIA, but it is an instance of a regulated international, market-based solution of the type that will be required in many sectors if net zero is to be achieved (CORSIA, 2020). CORSIA looks good in theory but the practice may disappoint. Why? Because CORSIA is being run by UN institutional surrogates, and they have a poor track record of enforcement of climate agreements. CORSIA is a diplomatic compromise on offsets (the GHG value of which is yet to be gauged). Because of this, governments must be ready to consider additional pricing mechanisms and to step in should CORSIA fail to deliver reductions.

It is reasonable to demand that the richest and most fortunate among us bear the greatest burden. Frequent flyers should pay an escalating carbon price, i.e. the more they fly and the more they use business class flights, the more they should pay. Data show why frequent flyers should pay much more. Only 12 percent of American flyers accounted for a staggering 68 percent of trips in 2017 (see Figure 6.8). Fully 53 percent of Americans do not fly at all. The steepness of this curve is similar in

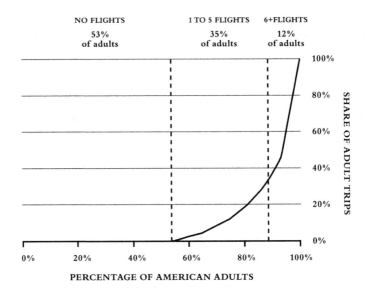

NOTE: Adapted from airlines.org/wp−content/uploads/2018/02/A4A−Air/TrabelSurvey−20Feb2018−FINAL.pdf. Numbers vary slightly from original due to rounding.

FIGURE 6.8 The rich fly the most

Source: International Council of Clean Transportation

Europe and practically vertical outside the advanced economies (ICCT, 2019). As the ICCT (2019) rightly states, 'Our global climate literally cannot tolerate widespread frequent flying, and people who currently fall into that group probably do need to modify their behavior'.

Those that pollute most should pay the greatest share and internalize the costs. Yet CORSIA fails to include the worst culprits: Corporate CEOs with their own planes. Signalling matters, and this needs to change. The most polluting individuals must pay for the privilege. If higher charges mean that corporate CEOs fly less, so be it. The result is a planetary plus.

What of technological developments? In this sector there are relatively few, and those that exist are in their infancy. At present, a demonstration flight of 30 minutes by a short-haul EV plane is the most notable progress, but there is no climate-friendly technology (other than hydrogen) or EV options for long-distance flights. Many of us just need to fly less.

Slowing and shifting shipping's route

Just as airlines and their customers must bear a larger burden via altered incentives, so too must shipping, which accounts for over 3 percent of GHG emissions. In a business-as-usual scenario, shipping emissions are projected to increase by 120 percent by 2050. If other sectors pursue GHG reductions in line with the Paris Agreement and shipping does not, this would mean shipping could amount to 10 percent of emissions by that date (Transport and Environment, 2020). However, international shipping, which carries roughly 90 percent of world trade, is essential to the global economy and is among the most efficient means of travel. Air is the most polluting, with 0.8063 kilograms of CO_2 per nautical mile; trucking comes next, at 0.1693 kilograms of CO_2 per nautical mile; and this is followed by rail at 0.1048 kilograms of CO_2 per nautical mile. Sea freight is the most efficient at 0.0403 kilograms of CO_2 per nautical mile. In 2020, moving cargo by ship was 20 times more efficient than by air, 4 times more efficient than by truck, and 2.6 times more efficient than by train.

Shipping is included in the Paris Agreement, and a series of steps are being taken towards net-zero goals. The International Maritime Organization (IMO) has set targets to achieve these goals. The 2018 IMO GHG strategy for shipping would require the shipping sector to reduce its emissions by at least 50 percent by 2050 compared to 2008. The strategy calls for the carbon intensity of international shipping to decline as an average across international shipping by at least 40 percent by 2030, and by at least 50 percent by 2050, compared to 2008.

This is a heavy lift. Major changes to fleets, operations, efficiency, and shipping speed will be needed. Governments and regulators should mandate disclosure, reporting of efficiency metrics, and the publishing of fleet targets and strategies to achieve GHG goals. Regulators should also consider mandating updating fleet efficiency, i.e. encouraging the scrapping of old ships (EU Commission, 2021). Regulatory mandates can change incentives, reporting, goals, and transparency and can speed diffusion of new technologies.

In 2020, much of the shipping industry remains inefficient, polluting, and uninterested in the necessary changes. Smaller shippers barely pay lip service to climate change. Governments need to act collectively to force the pace and set mandates and to ensure they are overseen by international regulatory bodies, including the IMO. Ultimately, penalties should be levied against polluting pirates. Only if this is done can market forces drive the desired results, punish laggards, and reward first movers, such as Maersk, the world's largest shipping line, which is steaming towards net zero (see Box 6.12).

BOX 6.12 MAERSK SAILS A NEW ROUTE

Maersk, the world's largest shipping line, has committed to net-zero emissions by 2050. To achieve this, the shipping line is investing in carbon-neutral fuels and ships. Ships have a 20- to 25-year lifespan. The company must have carbon-neutral ships commercially operating by 2030. It has invested US$1 billion since 2016 in pursing that goal. In addition to new vessels, the company has reduced its CO_2 emissions by 41 percent from its 2008 levels, and its goals are in line with the IMO strategy (Maersk, 2019).

Maersk has managed to decouple emissions from trade growth. Its ships have been cruising at slower speed, increasing efficiency and cutting GHG emissions. Maersk is now targeting a 60-percent reduction relative to cargo moved by 2030 from the 2008 baseline. Maersk's leadership recognizes that the climate change problem can only be solved by becoming carbon neutral. Efficiency will not be enough; they are leading the push for industry-wide solutions.

The firm is also testing sail technologies to increase efficiency and cut fuel use. Crucially, this technology can be retrofitted to parts of the existing fleet without sacrificing capacity, so they do not need to wait for new vessels to secure the GHG emissions benefits.

Innovative solutions already exist and include incremental design changes, solar-powered ships, and sail-assisted ships. Of those easily adopted, slowing down and adding sails offer the greatest improvements in ship efficiency (30 percent and 23 percent, respectively) and are simplest and least expensive. Overall, the pace of change, of renewal of the fleet and scrappage of the old fleet, must accelerate and reward those, like Maersk, that are taking a lead. Governments acting collectively must ensure shipping does not switch to underregulated pirate carriers flagged in lax jurisdictions. We cannot have another free-rider problem on the high seas. In shipping, constant innovation is necessary to steepen the diffusion curve, and this includes testing new fuels, such as ammonia (see Box 6.13).

BOX 6.13 A FOUL BUT PROMISING SHIPPING STENCH

MAN Energy Solutions, a multinational company based in Augsburg, Germany that produces large-bore diesel engines and turbomachinery for marine and stationary applications, is testing a ship engine design that runs on foul-smelling ammonia, a substance that may help replace bunker fuel in the mega cargo carriers of tomorrow. The technology will at first be built to handle both fuels and tested in that configuration in 2024.

Importantly, although less energy dense than bunker fuel, ammonia is denser than hydrogen, which is often touted as a possible replacement for oil, and it does not have to be stored at minus 234 degrees. Other shipping firms, including Eidesvik, are also exploring ammonia as a fuel.

Ammonia is not without its own challenges regarding handling, toxicity, and – when produced for fertilizer – GHG emissions. Here, too, innovation and R&D are essential. Researchers at Aarhus University and catalysts manufacturer Haldor Tospoe in Denmark are developing technology that can produce ammonia from water, air, and renewable electricity. As the researchers note, 'instead of utilizing fossil energy … we simply take wind and solar energy, and within minutes have liquid fuel at the other end' (BBC, 2020b). The technology will be commercially viable by 2022 or 2023.

A steepening glidepath before us all

It is not possible here to assess every sector, industry, and possible technology for improving GHG emissions. It would be an exercise in heating if not boiling the ocean, and climate change has already begun that. Nonetheless, the various examples above and differing governmental, sectoral, firm-level, and individual shifts underscore that the process of transformation and the acceleration of diffusion of new technologies has begun, even if it has a long way to go.

It should be clear from this brief account of several major sectors that across large parts of our national, regional, and global economy the climate change realignment response and redesign is unfinished, and it is barely begun in many areas. In 2021, getting to net zero is not a matter of incrementalism – of only small changes in existing practices. We risk triggering tipping points to dangerous new equilibria unless governments simultaneously support innovation, shift incentives, and speed the rates of new technology diffusion and adoption in numerous fields. Faster change is essential.

In 2021, countries from Europe to the US to China are increasingly crafting green industrial policies. This should be accelerated. Governments need to underpin green innovation and speed diffusion across their economies, shifting the signalling and incentives by doing so. In many sectors – from agriculture to industry to construction – breakthrough innovations are still required, and adoption of existing known and understood GHG-reducing technologies must occur.

A steeper S-curve is possible

Evidence shows a steepening of the S-curve is possible. Not only are we in a period of accelerated rates of technology diffusion generally – the IT-digitization-green revolution is all around us – but national examples indicate the rate of technology adoption is directly impacted by public policy, climate change regulatory actions, and anticipated changes, whether that is Sweden's carbon tax, Norway and the UK's combustion engine phase-out dates, or the Biden administration's climate change return to reality.

Rapid diffusion of green technologies is already underway and will continue to be an engine of economic growth. Denialist neo-Luddite assumptions about new technologies are wrong. Properly supported, regulated, and overseen new technologies have begun to power the green industrial revolution and that will continue, and in doing so they will begin to address problems of productivity and secular stagnation. The green revolution, sparked by renewables and much else besides, need not leave so many behind as we have in the digitized, neoliberal version of globalization.

The emerging Green Globalization 2.0 can be a markedly different and more equitable revolution. The green revolution need not be characterized by riches for the few and poverty for the many. Green Globalization 2.0 will come about through the interrelationship and nexus of government clarity on goals, support, shifting incentives, new regulations, and market and firm responses, rejuvenation, reimaging, reseeding, rebuilding, and redesign of entire sectors of our real economies. The leading firms across our economies are already seizing the commercial opportunities that this transition creates.

We see increasing evidence from leading-edge firms who are making the necessary climate change commitments, adjusting their strategies and investments, disinvesting, and profiting from their business model. These risk-taking firms are already reaping rewards from investors hungry for green assets. These firms and their leaders and employees understand that alignment and embedding of green and net-zero goals are essential for long-term economic success and profitability. They are assisting in the construction of Green Globalization 2.0 while increasing profits. This is not some fluttering slogan on a yellowing, forgotten office breakroom wall. Climate-aligned companies are undertaking wholesale redesign of approaches and mindsets. The leading firms are not just box ticking.

As for the laggards and denialists, they will be punished by investors. Many will go bankrupt, as Mark Carney (2019) observes. This is the nature of creative destruction in regulated, well-functioning market economies. Visible failures will increasingly mount as the rate of transition and the winners and losers in the new Green Globalization 2.0 are revealed.

We still have a long and difficult route ahead to secure Green Globalization 2.0. Success is not assured, but across our governments and economies actors increasingly understand what must be done. The economic transition is underway. It can be smooth, or it can be a more disjunctive, disruptive break. Which route we will travel depends on our governments, acting collectively and individually, their signalling, policy designs, and implementation of new technologies and industrial policies, and their speeding of diffusion rates.

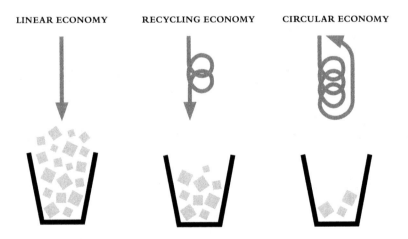

FIGURE 6.9 Time for a circular economic model

Source: Plan C Empowering Circular Features

Success depends on millions of realigned corporate decisions on how to invest, what to acquire, when to disinvest, and what to avoid. Success also requires us to ultimately adopt new understandings, stories, and models of how our economies, societies, and ecosystems will operate. In making this transition, we will move from a linear economy model, beyond a recycling economy model, to a circular economy model in which we recognize planetary limits and operate within them, while also ensuring economic sustainability, resilience, and equity (see Figure 6.9).

References

Acadia. (2019) 'The regional greenhouse gas initiative: 10 years in review' [Online]. Available at: https://acadiacenter.org/wp-content/uploads/2019/09/Acadia-Center_RGGI_10-Years-in-Review_2019-09-17.pdf (accessed: 22 September 2020).

AWEA (American Wind Energy Association). (2020) 'Wind energy in the United States' [Online]. Available at: www.awea.org/wind-101/basics-of-wind-energy/wind-facts-at-a-glance (accessed: 20 October 2020).

BBC. (2020a) 'Beer and crisps used to help tackle climate change', 7 December [Online]. Available at: www.bbc.com/news/science-environment-55207597 (accessed: 28 December 2020).

———. (2020b) 'The foul-smelling fuel that could power big ships', 6 November [Online]. Available at: www.bbc.co.uk/news/business-54511743 (accessed: 27 May 2021).

———. (2020c) 'McDonalds among food firms urging tougher deforestation rules', 5 October [Online]. Available at: www.bbc.com/news/business-54408544 (accessed: 28 December 2020).

Beerling, D.J., Kantzas, E.P., Lomas, M.R., Wade, P., Eufrasio, R.M., Renforth, P., *et al.* (2020) 'Potential for large-scale CO_2 removal via enhanced rock weathering with croplands'. *Nature*, 583: 242–248 [Online]. Available at: https://doi.org/10.1038/s41586-020-2448-9 (accessed: 29 December 2020).

BloombergNEF. (2019) 'New energy outlook 2019', 20 August [Online]. Available at: www.gihub.org/resources/publications/bnef-new-energy-outlook-2019/ (accessed: 24 January 2021).

Boer, J., Schösler, H., and Boersema, J.J. (2013) 'Climate change and meat eating: An inconvenient couple?' *Journal of Environmental Psychology*, 33: 1–8 [Online]. Available at: www.sciencedirect.com/science/article/abs/pii/S0272494412000618 (accessed: 28 December 2020).

BP. (2019) 'BP Energy Outlook 2019' [Online]. Available at: www.bp.com/en/global/corporate/news-and-insights/press-releases/bp-energy-outlook-2019.html (accessed: 24 January 2021).

Carbon Tracker. (2020) 'Decline and fall: The size and vulnerability of the fossil fuel system' [Online]. Available at: https://carbontracker.org/reports/decline-and-fall/ (accessed: 29 January 2021).

Carney, M. (2019) 'BoE's Carney warns of bankruptcy for firms that ignore climate change'. *Reuters*, 31 July [Online]. Available at: www.reuters.com/article/us-britain-boe-carney/boes-carney-warns-of-bankruptcy-for-firms-that-ignore-climate-change-idUSKCN1UQ28K (accessed: 29 December 2020).

Castellanos, J.D.A., Rajan, H.D.V., Rohde, A.-K., Denhof, D., Freitag, M. *et al.* (2019). 'Design and simulation of a control algorithm for peak-load shaving using vehicle to grid technology'. *SN Applied Sciences*, 1 August [Online]. Available at: https://doi.org/10.1007/s42452-019-0999-x (accessed: 29 September 2020).

CCC (Climate Change Committee). (2019) 'Progress in preparing for climate change: 2019 progress report to parliament' [Online]. Available at: www.theccc.org. uk/publication/progress-in-preparing-for-climate-change-2019-progress-report-to-parliament (accessed: 27 May 2021).

CEC. (2020) cec.org [Online]. Available at: www.cec.org.cn/guihuayutongji/tongjxinxi/niandushuju/2020-01-21/197077.html (accessed: 20 October 2020).

Cemex. (2020) 'Cemex announced ambitious strategy to address climate change', 19 February [Online]. Available at: www.cemexusa.com/-/cemex-announces-ambitious-strategy-to-address-climate-change (accessed: 7 October 2020).

China NREC (National Renewable Energy Center). (2019) 'China renewable energy outlook 2019' [Online]. Available at: http://boostre.cnrec.org.cn/wp-content/uploads/2020/03/CREO-2019-EN-Final-20200418.pdf (accessed: 20 October 2020).

Cole, C., Duxbury, J., Freney, J. *et al.* (1997) 'Global estimates of potential mitigation of greenhouse gas emissions by agriculture'. *Nutrient Cycling in Agroecosystems* 49: 221–228 [Online]. Available at: https://doi.org/10.1023/A:100973171134.

Cordis. (2020) 'CO_2 capture from cement production' [Online]. Available at: https://cordis.europa.eu/project/id/641185 (accessed: 29 December 2020).

CORSIA. (2020) 'Carbon offsetting and reduction scheme for international aviation (CORSIA)' [Online]. Available at: www.icao.int/environmental-protection/CORSIA/Pages/default.aspx (accessed: 29 January 2021).

Earth.com. (2019) 'Europe is phasing out coal faster than ever before', 29 October [Online]. Available at: www.earth.com/news/europe-coal (accessed: 29 January 2021).

EU Commission. (2021) 'Future of the Common Agricultural Policy' [Online]. Available at: https://ec.europa.eu/info/food-farming-fisheries/key-policies/common-agricultural-policy/future-cap_en#highergreenambitions (accessed: 29 January 2021).

Faraday Insights. (2020) 'Lithium, cobalt and nickel: The gold rush of the 20th century', Issue 6, Update June 2020 [Online]. Available at: https://faraday.ac.uk/wp-content/uploads/2020/06/Faraday_Insights_6_Updated.pdf (accessed: 20 October 2020).

Forbes. (2019) 'Renewable energy job boom creates economic opportunity as coal industry slumps', 22 April [Online]. Available at: www.forbes.com/sites/energyinnovation/2019/04/22/renewable-energy-job-boom-creating-economic-opportunity-as-coal-industry-slumps/?sh=1d83a7e33665 (accessed: 28 January 2021).

Future Farming. (2020) 'John Deere: We believe in electric tractors. 100%', 12 March [Online]. Available at: www.futurefarming.com/Machinery/Articles/2020/3/John-Deere-We-believe-in-electric-tractors-100-552869E/ (accessed: 23 January 2021).

Geroski, P. (1999) 'Models of technology diffusion', Paper 2146, Centre for Economic Policy Research [Online]. Available at: https://repec.cepr.org/repec/cpr/ceprdp/DP2146.pdf (accessed: 20 January 2021).

The Guardian. (2016) 'Making concrete green: Reinventing the world's most used synthetic material', 4 March [Online]. Available at: www.theguardian.com/sustainable-business/2016/mar/04/making-concrete-green-reinventing-the-worlds-most-used-synthetic-material (accessed: 8 October 2020).

GWEC. (2021) 'A gust of growth in China makes 2020 a record year for wind energy', 21 January [Online]. Available at: https://gwec.net/a-gust-of-growth-in-china-makes-2020-a-record-year-for-wind-energy (accessed: 24 May 2021).

G30 (Group of Thirty). (2018) *Banking Conduct and Culture: A Permanent Mindset Change.* Washington, DC: Group of Thirty.

———. (2020) *Mainstreaming the Transition to a Net-Zero Economy.* Washington, DC: Group of Thirty.

Iberdrola. (2021) 'Iberdrola group's climate commitment' [Online]. Available at: www. iberdrola.com/sustainability/against-climate-change (accessed: 29 January 2021)

ICCT (International Council on Clean Transportation). (2019) 'Should you be ashamed of flying? Probably not', 23 September [Online]. Available at: https://theicct.org/blog/staff/should-you-be-ashamed-flying-probably-not (accessed: 29 January 2021).

IEA (International Energy Agency). (2018) 'World energy outlook 2018', November. Available at: www.iea.org/reports/world-energy-outlook-2018 (accessed: 24 January 2021).

———. (2020) 'Global electric vehicle outlook 2020', June [Online]. Available at: www.iea.org/reports/global-ev-outlook-2020 (accessed: 20 October 2020).

IEEFA (Institute for Energy Economics and Financial Analysis). (2020) 'US coal companies battered by investors in 2019, as leading sector index drops 53%', 2 January [Online]. Available at: https://ieefa.org/u-s-coal-companies-battered-by-investors-in-2019-leading-sector-index-drops-53 (accessed: 20 October 2020).

IEEP (Institute for European Environmental Policy). (2019) 'NetZero agriculture in 2050: How to get there', February [Online]. Available at: https://ieep.eu/uploads/articles/attachments/eeac4853-3629-4793-9e7b-2df5c156afd3/IEEP_NZ2050_Agriculture_report_screen.pdf?v=63718575577 (accessed: 17 October 2020).

IMF (International Monetary Fund). (2019) 'Global fossil fuel subsidies remain large: An update based on country-level estimates', IMF Working Paper WP/19/89 [Online]. Available at: https://www.imf.org/en/Publications/WP/Issues/2019/05/02/Global-Fossil-Fuel-Subsidies-Remain-Large-An-Update-Based-on-Country-Level-Estimates-46509.

IPCC (Intergovernmental Panel on Climate Change). (2018) 'Special report: Global warming of 1.5 °C' [Online]. Available at: www.ipcc.ch/sr15 (accessed: 29 December 2020).

———. (2019) 'Special report: Climate change and land' [Online]. Available at: www.ipcc.ch/srccl/chapter/summary-for-policymakers/ (accessed: 17 October 2020).

IRENA (International Renewable Energy Agency). (2019) 'Global energy transformation: A roadmap to 2050' [Online]. Available at: www.irena.org/publications/2019/Apr/Global-energy-transformation-A-roadmap-to-2050-2019Edition (accessed: 24 January 2021).

———. (2020) Renewable energy statistics [Online]. Available at: www.irena.org/-/media/Files/IRENA/Agency/Publication/2020/Jul/IRENA_Renewable_Energy_Statistics_2020.pdf (accessed: 20 October 2020).

Kortenhorst, J. (2019) 'New report suggests the speed of the energy transition is rapid', 11 September [Online]. Available at: https://rmi.org/new-report-suggests-the-speed-of-the-energy-transition-is-rapid (accessed: 16 October 2020).

Linquist, B.A., Adviento-Borbea, M.A., Pittelkowa, C.M., van Kessel, C., and van Groenigenb, K.J. (2012) 'Fertilizer management practices and greenhouse gas emissions from rice systems: A quantitative review and analysis'. *Field Crop Research*, 135: 10–21 [Online]. Available at: https://linquist.ucdavis.edu/sites/g/files/dgvnsk6581/files/inline-files/2012-Linquist-et-al-FCR-Review-GHG-fert.pdf (accessed: 29 December 2020).

Maersk. (2019) 'Towards a net zero carbon future', 26 June [Online]. Available at: www.maersk.com/news/articles/2019/06/26/towards-a-zero-carbon-future (accessed: 29 January 2021).

Mazucatto, M. (2015) *The Entrepreneurial State: Debunking Public vs. Private Sector Myth.* London: Anthem Press.

McKinsey. (2020) 'The road ahead for e-mobility', 27 January [Online]. Available at: www.mckinsey.com/industries/automotive-and-assembly/our-insights/the-road-ahead-for-e-mobility (accessed: 20 October 2020).

Milner, H. and Solstad, S.U. (2018) 'Technology diffusion and the international system'. Draft paper [Online]. Available at: https://scholar.princeton.edu/sites/default/files/hvmilner/files/technology-diffusion.pdf (accessed: 16 October 2020).

New York Times. (2020) 'Long concealed records show Trump's chronic losses and years of tax avoidance', 27 September [Online]. Available at: www.nytimes.com/interactive/2020/09/27/us/donald-trump-taxes.html (accessed: 29 January 2021).

oilprice.com. (2020) 'Five energy stocks to buy before Christmas' [Online]. Available at: https://oilprice.com/Energy/Energy-General/Five-Energy-Stocks-To-Buy-Before-Christmas.html (accessed: 29 January 2021).

PepsiCo. (2021) 'PepsiCo doubles down on climate goal and pledges net-zero emissions by 2040' [Online]. Available at: www.pepsico.com/news/press-release/pepsico-doubles-down-on-climate-goal-and-pledges-net-zero-emissions-by-204001142021 (accessed: 29 January 2021).

Pickering N. K., Chagunda, M.G., Banos, G., Mrode, R., McEwan, J.C., and Wall, E. (2015) 'Genetic parameters for predicted methane production and laser methane detector measurements'. *Journal of Animal Science*, 93 (1): 11–20 [Online]. Available at: https://doi.org/10.2527/jas.2014-8302 (accessed: 29 January 2021).

PV Magazine. (2020a) 'Distributed storage could save Texas $344 million per year by deferring transmission and distribution costs', 11 May [Online]. Available at: https://pv-magazine-usa.com/2020/05/11/distributed-storage-could-save-texas-344-million-per-year-by-deferring-transmission-and-distribution-costs/ (accessed: 29 September 2020).

———. (2020b) 'LCOE from large scale PV fell 4% to $50 per megawatt-hour in six months', 30 April [Online]. Available at: www.pv-magazine.com/2020/04/30/lcoe-from-large-scale-pv-fell-4-to-50-per-megawatt-hour-in-six-months (accessed: 7 October 2020).

PVPS. (2019) 'Trends in 2019 photovoltaic applications' [Online]. Available at: https://iea-pvps.org/wp-content/uploads/2019/12/Press_Release_-_IEA-PVPS_T1_Trends_2019-1.pdf (accessed: 20 October 2020).

Qin, X., Li, Y., Wang, H., Li, J., Wan, Y., Gao, Q., Liao, Y., and Fan, M. (2015) 'Effect of rice cultivars on yield-scaled methane emissions in a double rice field in South China'. *Journal of Integrative Environmental Sciences*, 12: 47–66 [Online]. Available at: https://doi.org/10.1080/1943815X.2015.1118388 (accessed: 29 December 2020).

Rios-Nunez, S.M., and Coq-Huelva, D. (2014) 'The transformation of the Spanish livestock system in the second and third food regimes'. *Journal of Agrarian Change*, 5 (4): 519–540 [Online]. Available at: https://doi.org/10.1111/joac.12088 (accessed: 28 December 2020).

Roberts, D. (2020) 'We have to accelerate clean energy innovation to curb the climate crisis. here's how'. *Vox*, 16 September [Online]. Available at: www.vox.com/energy-and-environment/21426920/climate-change-renewable-energy-solar-wind-innovation-green-new-deal (accessed: 28 December 2020).

ScienceAlert. (2019) 'Scotland is now generating so much wind energy, it could power two Scotlands', 17 July [Online] Available at: www.sciencealert.com/scotland-s-wind-turbines-are-now-generating-double-what-its-residents-need (accessed: 20 October 2020).

Shiller, R. (2010) *Irrational Exuberance.* Princeton: Princeton University Press.

Springmann, M., Godfray, H.C.J., Rayner, M., and Peter Scarborough, P. (2016) 'Analysis and valuation of the health and climate change cobenefits of dietary change'. *Proceedings of the National Institute of Science*, 113 (15): 4146–4151, 12 April [Online]. Available at: www.pnas.org/content/pnas/113/15/4146.full.pdf (accessed: 28 December 2020).

Stiglitz, J. (2019) 'The climate crisis is our third world war: It needs a bold response'. *The Guardian*, 4 June [Online]. Available at: www.theguardian.com/commentisfree/2019/jun/04/climate-change-world-war-iii-green-new-deal (accessed: 28 December 2020).

Transport and Environment. (2020) 'Shipping and the environment' [Online]. Available at: www.transportenvironment.org/what-we-do/shipping-and-environment (accessed: 29 January 2021).

UN. (2020) '2020 is a pivotal year for climate – UN Chief and COP26 President', 9 March [Online]. Available at: https://unfccc.int/news/2020-is-a-pivotal-year-for-climate-un-chief-and-cop26-president (accessed: 30 September 2020).

UN GlobalABC.org. (2021) [Online]. Available at: https://globalabc.org (accessed: 29 January 2021).

Veugelers, R. (2016) 'Empowering the green innovation machine'. *Intereconomics Review of European Economic Policy*, 51 (4): 205–208 [Online]. Available at: www.intereconomics.eu/contents/year/2016/number/4/article/empowering-the-green-innovation-machine.html (accessed: 28 December 2020).

Wall Street Journal. (2020) '2020 was one of the worst-ever years for oil write-downs', 27 December [Online]. Available at: www.wsj.com/articles/2020-was-one-of-the-worst-ever-years-for-oil-write-downs-11609077600 (accessed: 28 December 2020).

WEF (World Economic Forum). (2019) 'The speed of the energy transition gradual or rapid change?', September [Online]. Available at: www3.weforum.org/docs/WEF_the_speed_of_the_energy_transition.pdf (accessed: 29 January 2021).

7

GREENING OUR STORIES INTERNATIONALLY, NATIONALLY, AND ESPECIALLY LOCALLY

> Narratives constitute reality as we know it by making sense of observations, leading us to new inferences, and models for a path forward.
>
> *Veland et al. 2018: 42*

> An economic narrative is a contagious story that has the potential to change how people make economic decisions.
>
> *Shiller, 2019: 3*

All human societies rely on narratives. We tell stories to our children; we describe our world in stories; we sing stories to one another. We paint visions and narratives on the walls of caves. Michelangelo painted creation and Bible stories on the ceiling of the Sistine Chapel in Rome, just as his great rival, Leonardo da Vinci, painted the narrative of the Last Supper on the refectory of the Convent of Santa Maria delle Grazie in Milan. Humans live through the stories they understand. They deconstruct complexity by building tales to explain the apparently inexplicable and confounding. Our narratives and stories help us make sense of things, draw conclusions, and move forward (Veland *et al.*, 2018). Economic narratives also change what we think, how we react, and the choices we make (Shiller, 2019).

When we humans change the nature of our conversations, we change our own stories, how we think about a subject, how we react, and how we plan. Changing arguments and narratives matters because stories fuel our responses, shift our actions, and adjust our understandings. We know that if we latch onto an erroneous story, or construct one based on nonsense, the outcome can be terrible both individually and collectively.

It is my contention in this book that the narratives we use – the tales we tell one another (economic, social, and planetary) – help dictate events and outcomes by sustaining viewpoints and by speeding or slowing shifts in community

DOI: 10.4324/9781003037088-8

understanding of common challenges, problems, and actions and the consensus that might be achieved on them.

Our stories can be economic and invariably include models of how we believe the world works. The stories integrate and include scientific facts or fictions to our benefit or detriment. Stories can be political and cultural. Our narratives are affected by crises and shocks, such as financial booms fuelled by 'irrational exuberance' (Shiller, 2000), or by sudden disasters, wars, or pandemics. The Covid-19 pandemic has visibly affected our common stories, shifting our understandings, and altered the boundaries of what we expect from one another and from our societies and governments.

How we respond as societies, communities, and individuals is affected by the narratives we believe and the events we experience, which operate as feedback loops. When we adopt altered narratives and ways of thinking about complex issues, the boundaries of what is acceptable, possible, and achievable also shift. Major events and crises alter what we consider possible and necessary. Covid-19 demonstrated this, enlarging what we discuss and agree is appropriate for government action (massive fiscal intervention, furlough schemes, etc.), what is necessary for our societies (social distancing and isolation), and how we act as individuals (what burdens we would bear and could or would do for others).

This is also what we see when we seek to shift the community consensus on climate change.

When the community consensus rejects climate change facts and adopts a counterfactual stance, it will resist action and stymie responses. Yet when a community is shocked by severe weather events and recognizes the facts of climate change and humanity's role, a denialist consensus can be shaken (as we saw in Australia). Real world events thus alter positions and open new understandings, narratives, and political accommodations. Communities can then accept alternative economic possibilities, environmental goals, and glidepaths.

In constructing a decarbonized world, 'a coherent strategy and a compelling strategic narrative are essential to closing the climate action gap on climate change' (Bushnell, Workman, and Colley, 2016: 2). Many communities need to speed a reimagining and dialogue, to help create new climate change constructs in our minds about the world we live in, our role in it, and our responsibilities, collective and individual.

New narratives create new realities

Countries, communities, and localities are in a constant, ongoing process of narrative evolution around how they discuss climate change, net-zero plans, and possible solutions. Such stories are specific to location and need to be fact-based, considered, inclusive, impactful, economic, and emotional. Precisely how a community responds to the climate transition will differ from place to place.

The communication of climate change narratives must match the policy urgency and be underpinned by the science and data. The messaging must be consistent, clear, and understood. Communities need to understand that climate change is not something happening 'over there'. It is happening right here, right now, hence the need for dialogue and discussion based on facts. When responsibility to act is realized, stories can mobilize others into action for social change (Reisman, 2008).

Debate and ownership

Addressing climate change requires devolution of the discussion, power, and decision-making authority to localities so that once there is an understanding of the scale of the climate change challenge, communities can consider what steps to take.

There are examples of how this devolution can work and how discussions can unfold and positively affect climate change dialogue. In 2021, this devolution of the debate and the construction of the localized solutions is patchy, i.e. it is stronger in some regions and locations (in some parts of Europe, for instance) and a great deal weaker in others. The debate, discussion, and devolution of solutions must accelerate across all our communities and be properly supported and resourced by increased levels of transfers from advanced to emerging and lower-income states if we are to change our climate change narratives and the outcomes.

Common understanding is essential and possible

The task – recreating commonalities of understanding in fractured societies, such as in parts of the US body politic – is especially urgent. The task of creating new climate change stories and narratives is not easy, but it is not impossible, either. We can see it happening in some communities and cities. We see it in the youth-driven climate change narrative that caught flight propelled by activist Greta Thunberg and her school strikes for climate, which seized the attention of the young people of the world in 2019 and 2020. We see it in the work of the Extinction Rebellion. In 2020 and 2021, we can see evidence of a narrative shift underway in more and more countries and communities. We have examples of how productive civil and civic conversations and dialogues can successfully occur.

Regions and cities must lead the way

As the ways in which we describe and discuss climate affect our collective climate change stories, our policy actions should also be local, relevant, relatable, and devolved. After all, we can implement and impact climate change goals more easily and tangibly at home. Confronted by possible disaster on a global scale, we can feel overwhelmed. Yet looked at on a local scale, possible adaptions, mitigations, and changes in economic and individual behaviours are more tangible and achievable. In our own communities, we can feel and breathe in the air improvements, help craft new, innovative solutions, and implement them across our towns and cities.

Despite the positive possibilities of local action on climate change, too often concerted effort to change the conversation on climate can run up against false narrative constructs that explicitly stymie action.

False narratives and tackling climate change

If we believe climate change is a hoax, we avoid seeing what is right before our eyes: Wildfires in California, record temperatures across the globe, and floods inundating our neighbourhoods. Thus, 2020 saw then-President Trump peddling a false narrative on climate change. He continued to deny climate change was real, even suggesting California could address the wildfires by sweeping its forests more often. Unfortunately, Trump was just the most prominent actor in a long line of climate change denialists among US leaders. America's history of false narratives around climate change is long and painful and could itself take up an entire book. Let me try and condense the history of distortion into a few hundred words.

The Luntz memo's deleterious effect

Until the early 2000s, global warming (now called climate change) was not a political issue in the US in the sense of there being a sharp divide between left and right. Voters cared about the environment as hunters, walkers, or environmentalists, and they loved their wild outdoors, national parks, and vast wilderness areas. The subject of global warming was rising in importance for some voters, who wanted action to address it. Republicans were worried that then-president George W. Bush was electorally vulnerable. Frank Luntz, the leading Republican Party pollster at the time, responded by writing a now-infamous memo in which he advised the president to counter the narrative by saying that the scientific debate remained open and a subject of controversy and contention (which, in fact, it was not). As Luntz observed:

> Should the public come to believe the scientific issues are settled, their views about global warming will change accordingly. Therefore, **you need to continue to make the lack of scientific certainty a primary issue in the debate**. [Original emphasis in the memo]
>
> *Luntz, 2002: 137*

Luntz understood that avoiding action on climate change required the public to believe that the science was still in question and the issue undecided, for once the science was clear, Americans would demand action from their leaders.

The last 20 years of America's destructive and backwards-looking debate, and its denialism, doubt, and faux disputes over the facts and the visible reality of climate change, can all be traced back to implementation of this effective but damaging strategy: An undermining of the public's understanding about the scientific consensus on climate change. Indeed, we heard the same narrative from Judge Amy

Coney Barrett during her October 2020 US Supreme Court confirmation hearing when she described climate change as, 'a very contentious matter of public debate' (BBC, 2020). Climate change is not contentious. It is a physical, empirical, observable fact (Carney, 2020). Yet Justice Barrett thinks otherwise, having learnt so from two decades of repetition of the false Luntz narrative.

Beware the danger of false narratives and rabbit holes

The Luntz memo shows that if we latch onto false narratives, they can give us a sense we understand what is really going on – but if that storyline is not based on facts, we can disappear down a rabbit hole into an Alice-in-Wonderland world of alternative reality.

Take the example of the wide-ranging QAnon conspiracy theory. This complex, multilayered conspiracy posits that elites in Hollywood, the government, and the media (generally liberals and Democrats) are secretly engaging in large-scale child trafficking and abuse. The narrative is fuelled by anonymous cryptic messages from a mysterious 'Q' who leaves 'crumbs' for believers to follow and decode. This story is garbage and nonsense. Yet a Pew poll conducted in the US in 2020 found that 20 percent of the people who had heard of QAnon thought it was 'somewhat' or 'very good for the country' (Pew Research Center, 2020). Axios polling found one-third of Americans open to the conspiracy theory, with views strongest in the American South, where between 12 and 14 percent of voters in Mississippi and Louisiana self-identified as believers (Axios, 2020). Adherents to the QAnon narrative are not harmless: Believers are classified as a potential terrorist threat in the US (Business Insider, 2019). On January 6, 2021, many of the rioters, criminals, and armed terrorists who stormed the US Capitol in an attempted coup were self-identified QAnon believers.

Lest you think a penchant for wild narratives is exclusively an American phenomenon, it is not. Variants of the QAnon false narrative have been picked up and are being spread in the UK, Germany, and elsewhere. In 2020, polling in the UK found a quarter of those polled believed in certain aspects of QAnon theories, such as the baseless idea that Covid-19 was released on purpose, an idea debunked by a team of medical experts from the World Health Organization during their investigation in Wuhan in February 2021. UK QAnon believers were visible at anti-lockdown protests across the UK in 2020 (Vice, 2020). QAnon variants are also growing in popularity among far-right adherents in Germany (Bloomberg 2020). Andreas Kluth warns: 'There's a vector of contagion that scares me more than SARS-CoV-2, MERS, Ebola, Zika or any other virus circulating nowadays. It's the spread of pathological memes, also known by their more traditional name: conspiracy theories' (Bloomberg, 2020).

Social media and amoral, unthinking algorithms

Efforts to reimagine climate change narratives can be undermined by the rapidly evolving social media landscape that militates against the emergence of reality-based

FIGURE 7.1 The spiral rabbit holes of social media separation and isolation

societal understandings of common problems that require urgent solutions. How and what we consume as media appears to be corroding commonality, destroying widely understood truths, blurring what is fact from what is fiction, and allowing the creation of 'alternative facts' (Conway, 2017). Voters then construct separate understandings of events unmoored from the same factual landmass, adrift on a sea of memes, Facebook posts, and propaganda. Social media tends to pit 'us' against 'them'; strips away polite, inclusive conversation and debate; and rewards trolling, shouting, invective, false stories, and ugly threads.

In 2021, the internet and social media have not so much democratized access to information as atomized our consumption of confirmation-biased, carefully AI-curated, increasingly alarmist stories that take us away from common debates and understandings of collective global and national problems. Screen-addicted readers and voters disappear down individually tailored machine-generated information and media rabbit holes, fuelled by state-directed misinformation campaigns, such as those fomented by Russia and others (see Figure 7.1).

Consumers are pulled along by unthinking, AI-designed, click-baited 'machine drift' (*New York Times*, 2020). We read reports algorithmically designed to cater to our existing views and further amplify and distort them, often in highly problematic directions.

In 2021, we do not read the same newspapers. Instead, we turn on our tablets and go directly to questionable sources. For instance, a 2019 survey found that more than half of American adults polled get their news from Facebook, making it the most popular social platform for news sourcing, with YouTube (!) and Twitter the second and third most popular at 28 percent and 17 percent, respectively, and a variety of other platforms such as Instagram, LinkedIn, Reddit, and Snapchat also making smaller but notable appearances (TechCrunch, 2019). These are dubious sources for life's decisions and discussions. Interested observers are worried. Observers polled by the BBC on the major challenges we face this century named the breakdown of trusted information sources as a significant threat to society (Pew Research Center, 2017). As Kevin Kelly, co-founder of *Wired* magazine, states:

> The major new challenge in reporting news is the new shape of truth. …
> Truth is no longer dictated by authorities, but is networked by peers. For

every fact there is a counterfact and all these counterfacts and facts look identical online, which is confusing to most people.

Anderson and Rainie, 2017

This might reasonably lead one to despair. Faced by such endless onslaughts of garbage information sources – fake news – how can we reach a common understanding on climate change and the path ahead, especially in countries – like the US – where the fact-based debate has been distorted for decades?

Talk about it: How conversations can change minds and advance the green consensus

An effective solution does exist. We need to stop looking at our screens and start talking civilly, locally, respectfully, and in person with one another again. We need to embed climate change dialogue within every community and on every level so we can all understand the reality of climate change and own it. We need to create new forums and new spaces where we can discuss, debate, and understand the climate change crisis so we can begin to plan our communities' future.

This not about selling green propaganda. This is about a recognition and an ownership of a physical planetary reality. Net zero is not a slogan – it is a scientific imperative (Carney, 2020). Conversations based on commonly understood and agreed facts need to be to be repeated and ongoing. Only by taking such civic, courteous debating steps can those of us in countries where climate change has been made ideological, depoliticize the debate and address the subject rationally and appropriately.

Is this type of conversation still possible? The QAnon example, the atomization of the public space, and the destruction of common understandings suggests perhaps not. In 2020, fully 48 percent of Americans voted for a climate denialist. Does this signal too many of us are increasingly unable to pull ourselves away from our separate screens and streams, and that broad and deep understanding is no longer possible?

I refuse to accept that positive dynamic reimaging – the creation of new, compelling, meaningful, face-based, and impactful stories – is impossible. Thankfully, there are working examples of how civic and civil debates can shift our stories and policy outcomes decisively and significantly in positive and important ways.

Let me start with a true story about sewerage, libraries, and taxation.

Sewerage, libraries, and taxation

Denver, Colorado has been a boom town since the early 2000s. The mile-high city on the edge of the Rocky Mountains is a tech and education hub fuelled by an influx of 4,000 to 5,000 new young workers a month, all wanting to balance work life with bombing down double black diamonds in Breckenridge or Aspen, or biking to Leadville along the highest and hardest, most oxygen-starved roads in America. Denver in 2020 was the fifth-fastest-growing city in America.

In the first decade of the twenty-first century, its services – its water and sewerage lines, schools, and other civic services – were being pushed to the breaking point by this rising population. The new mayor, John Hickenlooper (now the junior US Senator from Colorado, having been elected in 2020), knew he had to act. He needed to ask people to pay more taxes to sustain the city and ensure its liveability. Surely freedom-loving, gun-toting, tax-averse Americans would never vote for tax increases.

In 2007, Mayor Hickenlooper proposed a US$550 million bond issuance and property tax increase package named 'Better Denver'. Few believed Hickenlooper would be successful in getting it accepted; he was seeking to buck a longstanding refusal of Americans to pay for the services they demand. However, it turned out that reality, when transparently, carefully, and clearly discussed, repeatedly debated, and better understood, can change the narrative and outcomes. Hickenlooper spent a year going from library to library, public meeting to public meeting, describing the public services crunch the city was facing, the sewerage demands, the new plumbing required, the schools needed, the many other related projects the growing city desperately needed, and the money that had to be raised to sustain Denver's economic rise. My sister-in-law, a Denver resident, remembers attending these earnest discussions on sewerage, schools, roads, and taxes at her local library.

Thanks to careful, balanced, fact-based advocacy backed by various communities and stakeholder organizations, the conversation and consensus among city voters evolved. In the end, the city's electorate voted for the raft of bonds and tax increases. A middle-of-the-road Democratic mayor sold significant tax increases to a sceptical audience because he first laid out the facts and challenges openly and clearly and then made the case for action and expenditure repeatedly and calmly. His work was not exciting; rather, it was tiring and repetitive, but good government often is. Government involves open discussion of, and agreement about, possible solutions to difficult common issues. Sloganeering and point scoring it is not. Yet when you get it right, the consensus can change for good.

Hickenlooper's investment of time and effort, and his success, laid the foundation for more than a decade of expansion and economic growth, as Figure 7.2 shows. Note that Denver was barely impacted by the 2008–2009 global financial crisis. The essential and positive fact-based story Hickenlooper told, and the funds he then secured and the investments he made in the city's infrastructure worked and contributed to a booming city and economy.

Hickenlooper helped build the city we see in 2021. Talk to Denver residents today and there is little doubt that his success was the city's success, helping to ensure its continued economic growth and today's prosperity. What we witnessed in Denver in the 2000s was local voters understanding the facts, taking the long view, and being willing to pay more for a better future together. I believe we can achieve similar results through shifting the conversation, dialogue, and policy choices on climate change, and help support the transition and green technological revolution. Fact-based, calm, community conversations across political divides can succeed.

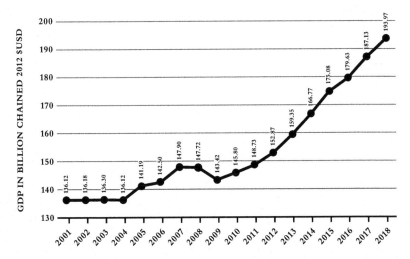

FIGURE 7.2 Denver on the rise – fuelled by straight talk and investment

Source: BEA © Statistics 2019.

Let's take another example of constructive conversations, one about retirement and pensions.

Pension demographics and longevity

Pensions is the so-called third rail of politics, for no one will touch it, however necessary adjustments might be. Yet in 2002, in Britain, London School of Economics professor Adair Turner and his colleagues succeeded where many others have failed. He convinced the British to work longer before retirement to keep the pensions system functioning and fair.

Pensions, or more specifically the societal cost and burden transfer across generations, is a huge challenge for all societies. As societies advance, demographic drivers coupled with increasing wealth lowers total factor fertility rates, shrinking the worker-to-retiree ratio. Gradually, fewer and fewer workers must support more and more retirees for longer and longer because simultaneous with economic growth, populations are living longer and longer past state retirement ages of 62 or 65. Few of us want to delay retirement, but, when first established, pensions retirement age calculations estimated most people would live one or two years past retirement. Yet in 2021, most of us will live 15 or 20 years past retirement. The average life expectancy for a child born today in developed economies is 90. Societies cannot pay for people to spend a third of their life in retirement. Such costs are rising and must be borne by existing and future workers, firms, and society. This cost – the difference between what people expect to receive and what workers and societies can provide (without retirement age and tax adjustments) amounts to a global US$$15.8 trillion pensions funding crisis (G30, 2018) across advanced economy countries.

Yet in 2021, because of Professor Turner, the pensions funding crisis has been averted in Britain. What did he do and why did it succeed? The Labour government created a pensions commission, chaired by Turner in 2002, which facilitated and participated in a national debate on the sustainability of the UK pensions system. The commission addressed the need to change benefits, taxes, and retirement ages gradually and progressively, in line with the longer lifespans. The commission published the Turner Report and recommendations in 2004, with a final report in 2006.

In the four years during which the commission worked, they took to the road and held pensions discussions across the country to explain the need for reform. In a measured and careful way, they presented nonpolitical facts and analyses of the challenge facing the UK and presented the options. The commission took hundreds of verbal and written testimonies from all stakeholders across the country. Their goal was to depoliticize the debate and build consensus. Professor Turner was clear that facts must drive decisions, stating:

> If you look at the first report of the Pensions Commission ... what you find is a voluminous fact base ... we were trying to make sure that people could not disagree on this analysis ... I think it's very useful in these processes ... to see if everybody agrees on the prognosis of what will happen if policy does not change.
>
> *Turner in IFG, 2007: 89*

The events held across the UK were each attended by around 300 people. Attendees were polled at the beginning of the discussion to see where they stood on the report's findings. The poll showed on average around 80 percent of participants were opposed to raising the state retirement age at the start of discussions. By the end of a day of discussions, having been taken through the facts and analysis, attitudes were fundamentally different, and participants were more open to change and more ready to act. It was hard work, but it paid huge dividends.

The commission ultimately issued the following recommendations: (1) create a low-cost, nationally funded pension savings scheme that individuals would all automatically be enrolled into, with an option for opting out; (2) make the system less means tested in order to minimize disincentives to saving financed partly by an increase in taxes devoted to pensions; and (3) re-link the basic state pension to average earnings growth financed in part by a steady increase in the state pension age designed to keep the proportion of life spent in retirement constant. All three steps were implemented by the government, putting UK pensions, public and private, on a stronger, more sustainable foundation. The commission had crucially, 'created space for measures that had previously been unthinkable' (IFG, 2007: 94).

This brief history of the UK Pensions Commissions demonstrates how an open and deliberative policy approach and dialogue can change the narrative story on a contentious issue, depoliticize it, change opinions, widen possible policy solutions, and support their implementation based on agreed facts.

The Denver and UK examples are the antithesis of politics by Facebook meme, propaganda, and disinformation. They represent considered, democratic, collegiate discourse in public for the public goods we all require. The key lesson from both examples is that facts matter. Presented clearly and discussed at length across all communities and stakeholders, facts can form the basis for a shift in consensus and open up policy outcomes hitherto thought impossible. It turns out that misinformation can be replaced by clarity, consensual collective action, and much better outcomes.

A further, current climate change conversation example sheds additional light on this type of inclusive public debate and process as applied today.

Scotland's big climate conversation

In 2019, the Scottish Parliament passed the Climate Change Act, legally binding the government to the climate change net-zero goal by 2045. The government clearly understood the size of the challenge, estimating that over 60 percent of measures to achieve net zero require at least some change in the way Scottish society operates. The government also knew dialogue and discussion would be essential to expand understanding of the common climate change challenges within the country. To that end, they initiated the Big Climate Conversation, a collaborative, inclusive, nationwide dialogue that would comprise a series of discussions with stakeholders across Scotland about how Scottish communities could tackle the climate emergency.

The dialogue took place in 125 workshops and community and digital events throughout the country as guided conversations led by facilitators. Participants were asked their views at the start, in writing or via the web, and again towards the end of the conversation.

The positive results of this countrywide collaborative conversation are clearly seen in the participants' self-reported level of knowledge about climate change before and after the conversations, as in Figure 7.3.

The results in Figure 7.3 show that 'climate conversations can be an effective tool for improving knowledge about climate change' (Scottish Government 2020: 17). Both the targeted and the open audience meetings led to an increase in understanding. The change was most marked among the targeted audience, which is explained by the make-up being more diverse and not an audience of self-selected, already partially informed attendees.

Scotland's Big Climate Conversation was designed to help support the achievement of the country's net-zero target by 2045, which is a 'step change in ambition' for the country, according to the UK Committee on Climate Change (BBC, 2019b). This aggressive target will likely be easier to secure because of this successful countrywide discussion, as it has helped create a broader understanding of what needs to be done.

Talking together changes our story and our outcomes

As I have said, we live, talk, and think by storytelling. The Denver tax conversation, the UK pensions debate, and the impactful Scottish Big Climate Conversation

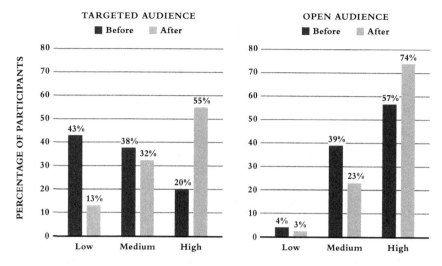

FIGURE 7.3 Understanding the facts on climate change alters perspectives

Source: Scottish Government 2020.

are examples. All three show us that we rely on stories to understand complex subjects, internalize lessons and takeaways, and alter our plans and actions. The best artists paint us a story. The best musicians sing us a story. The best policymakers and economists should address climate change with narratives and stories that are empirical and emotional, with a beginning, a middle, and an end, and not just rely on a model and an optimum price point.

Policymakers seeking to speed the transition to net zero need to foster and sustain interweaving conversations at multiple levels and in numerous forums, like these examples, that move and alter our stories around climate change, its effects, and our collective possible responses. These conversations can set the stage for communities' responses to climate change.

These examples of dialogue and debate have the following lessons for those seeking to reshape our stories to better address climate change and the necessary transition to net zero:

- *Leadership always matters.* The commitment of charismatic, credible leaders is essential. Leaders can articulate the facts and goals and diffuse opposition or convince recalcitrant opponents to consider change.
- *There is merit in separating the explicitly political from the climate change dialogue.* Conversations and dialogue cannot be an exercise in shouting slogans and abuse at one another. This is about reality and fact-based discourse, not shouting half-truths.
- *Lay out the facts first.* When a population does not fully understand the facts at issue, it is first essential to diagnose matters and lay out the data supporting the factual basis before considering possible prescriptions. With the facts on the table and generally understood, the options can then be discussed.

- *The process of discussion must be open.* The process must be broad and include all stakeholders, and the discussion must go into the necessary depth and treat all with respect. Open engagement is key when seeking to shift narratives on complex topics like climate change.
- *Changing narratives takes commitment and repetition.* It takes time and commitment to change the stories people use to understand difficult subjects. When confronting a difficult policy challenge, start now, as it will probably take longer than you think.
- *When a problem is multifaceted do not allow yourself to be viewed as beholden to any specific stakeholder or group.* Successful conversations do not presuppose one side or one solution over another; they need to be viewed as fair and reasonably deliberative processes, in order to arrive at reasoned, accepted, and supported conclusions.

Carefully handled, these types of climate change conversations and dialogue can and do change hearts and minds. The conversations are local, direct, civil, considered, and additive. They pull people away from their screens and away from online, aggressive anonymity and require people to treat one another as neighbours, friends, and fellow citizens. These types of dialogue can build inclusive understanding of common threats and the positive possibilities of renewal. We can have conversations where we understand one another, agree on the facts, and construct new climate change stories and green policy options for the future.

Create narratives that resonate with a locale and with your community

As we create these regreened climate change stories and narratives, they will be different from place to place, even as they have common factual elements. The story in the scorching deserts of Arizona will be different from the story in hurricane-hit Louisiana. The tales told in Greenland will be different from the tales of wildfires and species extinction in Australia. This is the nature of storytelling – to resonate, it must be local, real, and relatable. Climate change dialogues place facts into context to help us understand how climate change is relevant to our lives and what we can and should do about it. As Steger notes:

> While climate stories are individual perspectives, it is our collective stories that have the power to shift the narrative. ... Listening, compassion, and personal storytelling are tools that can connect us and nurture a common ground where true change can begin.
>
> *Steger in Climate Generation, 2020*

Collective wisdom on climate change has a greater momentum, a greater ability to trigger cooperative action. Hence the importance of a multiplicity of conversations and story making.

Narrative shifts can be slow until they are sudden

Often, an existing consensus will appear resistant to change until a critical point is reached, and then the narrative can be surprisingly abrupt. History shows us that a consensus shifts slowly – often imperceptibly – until reality or shocks shake the old assumptions loose, forcing the adoption of new understandings or stories. As Nelson Mandela reportedly said in a speech in 2013, 'It always seems impossible until it's done'. We know severe weather events can precipitate a climate change reappraisal, such as the wildfires in the Australian Outback. A charismatic leader can also personify, clarify, and alter the narrative and thus help speed the adoption of a new story and ending, as Mandela did for his nation.

The Greta effect

Perhaps no single individual has had more of an effect on how the next generation thinks about climate change than Greta Thunberg, a teenage activist who in 2019 started a one-person 'skolstrejk för klimatet' (school strike for climate) in front of the Swedish parliament, standing there alone every Friday. A movement was born out of Thunberg's anger, her charisma, her eloquence about what was at stake, and her visible distress at the failure of global leaders to address climate change 'right here, right now'. Thunberg's speech at the UN in 2019 catapulted her and the climate change emergency to global prominence among school children, young people, and adults. Her speech at the WEF on 25 January 2019 was a rhetorical slap in the face for corporate CEOs.

> Our house is on fire. According to the IPCC we are less than 12 years away from not being able to undo our mistakes. In that time unprecedented changes in all aspects of our society needs to have taken place, including a reduction in greenhouse gas emission by 50 percent. … On climate change we have to acknowledge we have failed. … But homo sapiens has not failed. We can still fix this.
>
> *Thunberg, 2019a*

Her school strike movement caught narrative fire in 2019 and she participated in demonstrations with millions of young people across Europe, with hundreds of protests seen across the globe that spring, summer, and fall involving approximately 6 million protestors. It was the largest youth movement on climate change the world has ever seen.

Thunberg travelled to the 2019 UN Climate Action Summit to address delegates. She travelled by sailboat from Sweden rather than fly because the carbon footprint matters. She knew leaders lead by what they say, but especially by what they do. Conduct signals what you really care about. At the UN summit, Thunberg castigated leaders for inaction, stating famously:

This is all wrong. I shouldn't be up here. I should be back in school on the other side of the ocean. Yet you all come to us young people for hope? How dare you! You have stolen my dreams and my childhood with your empty words. And yet I'm one of the lucky ones. People are suffering. People are dying. Entire ecosystems are collapsing. We are in the beginning of a mass extinction. And all you can talk about is money and fairytales of eternal economic growth. How dare you!

Thunberg, 2019b

Thunberg's remarkable impact on the climate change conversation can be seen in the colossal spike in global Google searches for her name at the peak of the school strike and on the day of her remarkable UN speech (see Figure 7.4).

A gradually rising level of concern over climate change can also be seen in the long-term search data from Google, as seen in Figure 7.5. You can see the spike in people trying to understand climate change in 2018 and especially in 2019 as the 'Greta effect' pushed climate change higher on people's list of global concerns.

Pollsters have confirmed a pronounced Greta effect, with young people taking up online activism as a result of her leadership. Ofcom, as the UK Office of Communications is known, found British children in 2019 were more likely to have used social media for activism purposes than in the previous year (Business Insider, 2020). This increased activism is visible in many countries including the US (NEA, 2019), where young people are demanding climate change action and are increasingly adopting a climate change narrative, spurred on by Thunberg. This narrative and activist action has not stopped because of the Covid-19 pandemic.

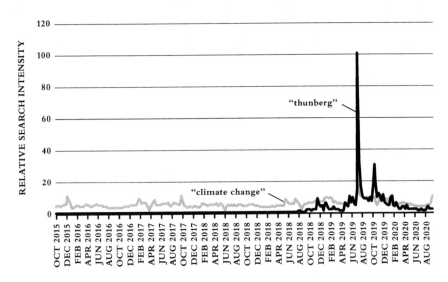

FIGURE 7.4 The Thunberg effect

Source: Google.

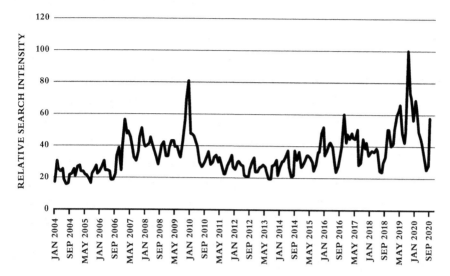

FIGURE 7.5 People's concern about climate change is growing

Source: Google.

The Fridays for Future campaign, which was started by Thunberg in August 2018 when she and other young activists sat in front of the Swedish parliament every school day for three weeks to protest the lack of action on the climate crisis, led to 3,500 strikes and marches across the world in September 2020 (Harvey, 2020), building on the 6 million people who participated in protests a year earlier, in 2019. This progressive, positive dynamic youth activism is filtering into electoral outcomes as these activists become motivated voters. They are:

> a surging cohort of [American] young voters who, for the first time in a presidential election, rank climate change as one of their top priorities. [They] orchestrated the largest global climate protests in half a century, and take credit for pushing environmental issues to the forefront of 2020 campaign.
>
> *Kaplan, 2020*

Today, young people are more likely than older voters to agree with the scientific consensus that the world is warming because of human activities and to demand action. For instance, a September 2020 poll in the US found 16 percent of respondents between ages 18 and 29 rated climate change as their most important issue; only the economy ranked higher among those polled. The saliency of climate change is rising globally. In 2021 the largest poll on climate change ever conducted included 550,000 people across 50 countries. The poll found 64 percent of participants saw climate change as an emergency (BBC, 2021). Voters can see climate change happening around them and are alarmed. They want action. Activism and new climate change narratives are taking root among other groups and this is affecting the climate narrative and politics in the UK and elsewhere.

Extinction Rebellion's direct effect

The Extinction Rebellion (XR) is arguably one of the most effective direct-action movements on climate change. XR is a decentralized, international, and nonpartisan movement using nonviolent direct action and civil disobedience to persuade governments to act justly on the climate and ecological emergency. XR activists blocked city centre roads in major cities across the world in 2019 and 2020 (including in London, New York, Washington, DC, and many other locations). XR has two key demands: (1) governments must tell the truth by declaring a climate and ecological emergency, working with other institutions to communicate the urgency for change; and (2) governments must act now to halt biodiversity loss and reduce GHG emissions to net zero by 2025. The XR movement takes the lessons of civil disobedience and applies them to the climate crisis. XR has had marked and rapid successes, for example, in the UK, where the parliament passed a motion announcing a 'climate emergency' in 2019. XR's language echoes Thunberg's warnings:

> We are standing on a precipice. We can acknowledge the truth of what we are facing, or we can continue to kick the can down the road. … We are already locked into a certain amount of warming and biodiversity loss, but there is still time to change this story.
>
> *Extinction Rebellion, 2020*

The school strike movement, Fridays for Future, led by Thunberg, the XR campaign, and other related movements put pressure on existing backward, destructive narratives and stances. As Angus Satow observed, 'We need groups like XR and the school strike movement to keep the pressure on and drive the climate crisis up the agenda' (Satow in Taylor, 2020). These youth and climate movements are derided and pilloried by those on the right, but Thunberg, XR, and others are altering our narratives as they challenge and critique existing stories and inaction. The groups show us narrative shift and dialogue is not only a committee process. It can also be a self-forming activist- and community-led dynamic. In the end, changing our stories on climate change and the transition is not either/or but both and more.

Climate change narratives are also spiritual

Our climate change narratives and dialogues are also religious and spiritual. A resonant spiritual voice is heard in Pope Francis' 2015 encyclical, 'Laudato Si', which deserves to be read and digested in full by anyone seeking a strong and emotive religious defence of the planet and our place on it. In the encyclical, the Pope describes climate change as having ethical and spiritual roots. The Pope urges the reader and his church not to look for solutions only in technology but also in a change of humanity and the moral orientation of the human heart – to replace consumption with sacrifice, greed with generosity, and to recognize the intrinsic value of the world and creation. The Pope is reaching for a form of politics and economics that can better serve the common planetary good. Thus:

> Politics must not be subject to the economy, nor should the economy be subject to the dictates of an efficiency-driven paradigm of technocracy. Today, in view of the common good, there is urgent need for politics and economics to enter into a frank dialogue in the service of life, especially human life.
>
> *Francis, 2015: 189*

The Pope's striking and forceful encyclical takes aim at rationalist, utility-focused, morally shrunken neoclassical economics:

> The principle of the maximization of profits, frequently isolated from other considerations, reflects a misunderstanding of the very concept of the economy. As long as production is increased, little concern is given to whether it is at the cost of future resources or the health of the environment; as long as the clearing of a forest increases production, no one calculates the losses entailed in the desertification of the land, the harm done to biodiversity or the increased pollution.
>
> *Francis, 2015: 195*

The Pope calls for a moral reappraisal, a recognition that we need one another, that we must act in a moral and ethical manner towards the environment. His is an intervention that is a comprehensive and sweeping religious vision and narrative on climate change, based on facts, supported by Christian spirituality, demanding action. Pope Francis' belief that the religious among us have a duty of care for the environment is echoed by many other faiths and leaders and is 'centered on the concept of "ecojustice," a comprehensive social and ecological vision of the interconnection of *all* of life' (Gottlieb, 2008).

Such multifaceted emotional, religious, and spiritual messages resonate with the messages of Thunberg and XR. This does not surprise me, since the drivers within narratives on climate change from leading voices are more often than not emotional and spiritual, asking us all to construct stories that take the scientific and the economic and connect them to our internal moral and ethical beliefs about emotional links with the world we live in, the woods we walk in, the beaches we stand on, and the rivers we fish.

In 2021, I believe the stories we construct about the planet and climate change are visibly and audibly shifting in response to multiple pressures, appeals, and conversations, both organized and spontaneous. We are beginning to construct new ways to foster the conversations needed to depoliticize climate change and address the net-zero transformation as a necessary, urgent planetary objective for us all.

Declaring an emergency shifts the narrative

Some peoples, countries, and communities are further down this narrative journey than others. As I have already observed, many European states and their cities, regions, and electorates have already adopted aspects of the climate change narrative. This increasingly involves declaring a climate emergency, with many cities, including Edinburgh (2019), London (2019), Oslo (2019), and Paris (2019), and many countries, including Austria (2019), Scotland (2019), and the UK (2019), taking this step. Words matter, and announcing an emergency has performative, dynamic galvanizing effects at many levels of government.

The statements are a declaration of a war on carbon. By declaring war, communities' collective narrative S-curves of policy innovation and transmission/adoption can steepen suddenly, underscoring again that narrative and public policy shifts often are slow, until a certain point is reached, and then change can occur very fast indeed.

Today, in 2021, we are all in a race against planetary climate tipping points and shifts to dangerous new equilibria. It is heartening to see evidence that similar types of narrative tipping points are now happening, which might lower the risk of reaching planetary climate tipping points of no return.

Figure 7.6 shows a jump in the number of governments declaring a climate emergency from almost zero in July 2018 to 1,814 in July 2020. These governments provide services and work for 830 million people in 30 countries. Of course, declaring an emergency is only the first step but doing so triggers widespread actions across government planning, forecasting, and service delivery. It can fundamentally alter the policy process and precipitate actions that cascade across communities, regions,

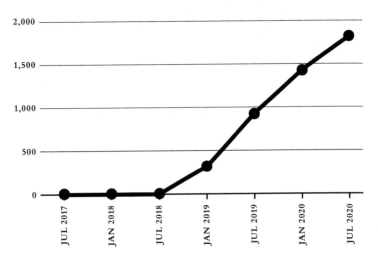

FIGURE 7.6 National, regional, and local governments declare climate emergencies

Source: Climate Mobilization.

cities, and countries. To be sure, these locations are still a minority of the world's population and political units, but the steepening narrative S-curve is clear.

Many states and regions are still too slow in altering their narratives and actions. Some have yet to start – and start they must. Nonetheless, the signals from this heightening sense of urgency in the last two-plus years is indicative of a rise in voters' demands for action on the climate and a concomitant rise in the importance of the climate emergency for national and local governments.

What of America? Can we count on US voters to change America's climate story and response? The Trump administration was a denialist negater of progress and a danger to the planet. Where do Americans stand on the climate change narrative today?

Americans are changing their climate story

The Greta effect, youth-driven evolving narratives, and other factors may be speeding up the rate of shift in views in the US. Americans in 2021 appear more ready to discuss climate crisis options and how to respond, locally and nationally, led by the Biden administration. It is still true that in America 'the climate conversation can sometimes feel like a shouting match in a roomful of children wearing earplugs' (Segal, 2019), but change is afoot.

It is possible that Americans can overcome internet-fuelled tribal currents and discuss climate change and the way forward. Reality may be triggering narrative shifts, forcing a realization of the danger and the need to discuss options and to act faster.

The end of the Luntz effect

Increasingly severe weather events, fires, flooding, and hurricanes are driving a change in views on climate change risk in America. By 2020, the corrosive Luntz effect (i.e. a denial of climate change scientific consensus and fact) on Americans was waning. Polls show that, in 2020, 72 percent of Americans believed climate change is happening. The number rises still more where the impacts are most visible and tangible. In the scorching panhandle of Texas, between 75 and 77 percent believe climate change is happening. In California, that number is higher still, at 78 to 81 percent. Once you believe climate change is real, you demand action, just as the Luntz memo concluded.

Crucially, in 2020, three out of four Americans polled supported regulating CO_2 as a pollutant. The number was higher still in some locations, such as Arizona (76 percent), California (79 percent), Massachusetts (82 percent), New Jersey (77 percent), and New York (83 percent) (Yale, 2020). Results like these suggest that Americans are well ahead of the Republican Party and its denial of climate change. American majorities now want presidential, national, congressional, and local action on climate change. This helps explain why candidate Biden's embracing of a major green industrial policy and massive investment programme as part of his election manifesto was not controversial in the 2020 presidential campaign. Trump tried to make it so, but it didn't work. When most voters want to tax carbon and believe climate change is happening, calling it 'a hoax' just underscores the denialist's ignorance.

In 2021, the Biden administration and Americans are ready for a fact-based, calmer dialogue and a climate change narrative shift. This will make the policy path to global net-zero goals easier and the possibility of success in COP26 on multiple fronts greater. Of course, there is still a great deal of talking to be done, conversations to be had, and listening to do. Nonetheless, as Segal argues, the climate conversation is not irreparably broken:

> We can't take away those unhelpful narratives. ... But we can add new ones. ... Scientific narratives, if they're done right, are some of the most powerful of all. They teach us more than facts, mechanisms, and procedures. ... Perhaps most important of all, they situate us in the world.
>
> *Segal, 2019*

The Biden administration policy shift supports a global climate break point

The Biden administration is signalling through its green policy goals and greening of industrial policy a step change in the US climate change policy stance. The administration is rejoining the Paris Agreement. It has announced a net-zero 2050 target. The Biden administration should now take a constructive leadership role, in collaboration with its long-time allies, on carbon pricing and the glidepath ahead.

With the change in US leadership, the policy possibilities shift and the opportunities enlarge. In 2021, the US is rejoining the international consensus and conversation on climate change goals and the paths ahead. America can turn its innovation and industrial might to the task of tackling climate change and harness markets to maximize the rate of change.

Announced and anticipated US policy shifts have already begun to positively impact market narratives. General Motors' announcement that it will phase out gas- and diesel-powered vehicles in 2035 is a stunning instance of the speed of change.

A combination of policy levers, altered expectations, and firm-level decisions will strengthen the sense of a tipping point among market participants, investors, and actors, as a critical mass of support for more dramatic action takes shape.

Looking ahead, in tandem with the urgent policy shift, the Biden administration should start planning for national conversations about climate change. Grounded in the data – i.e. in science and facts – these conversations can build on and cement the majority's recognition that climate change is happening. The conversations can help communities and cities to consider, based on agreed facts, what to do in response. Such conversations, supported by stakeholders in communities across America, will take time, but they are a necessary part of creating new climate stories across America.

Just as national governments, including the Biden administration, should engage in and support such national civic conversations, so should our local communities and leaders.

Devolution of the climate change dialogue and the ability to act

A devolution of the climate change dialogue and the power, authority, and resources to speed the transition are essential to the climate change response in all our communities. It is discussion, action, and planning at the local, devolved level that can often be more respectful, tangible, forward looking, and collective and less confrontational. It is in our regions, cities, and communities where many climate change policy responses and decisions can be felt in our daily lives. The UK shows us both how this can work and how not to do it.

The UK example

The UK is demonstrating both how devolution of power can work effectively and how it should not be approached. In Scotland, Wales, and Northern Ireland, devolved parliaments and assemblies have led on net-zero goals. The regions have committed to aggressive targets and started the climate change conversation across their communities. This process is consultative, representative, empowering, and inclusive, and is working well at the local level; communities, in discussing and debating climate change, are increasing their understanding about what is necessary for the green transition. In Scotland, we see the realization of credible, predictable policies, and how they have drawn forward market decisions, shifting incentives and market expectations and altering pricing decisions. This is the UK's devolved federal structure (created by former UK prime minister Tony Blair in the 1990s) paying climate change dividends. The devolved governments have been able to foster the needed climate change dialogue, support the narrative shift, and make it actual via policy.

Yet, the current UK government, headed by Prime Minister Boris Johnson, is pursuing a political and institutional policy of centralization of power. For other political economy reasons, the government aims to constrict devolution and pull authority back to London. This is the reverse of what is needed to engage and reconnect with local communities struggling with economic ills and climate change challenges. If governments want to foster dialogue, this needs to be local and inclusive. The communities must have the power, and the authority and resources, to make policy changes once a new narrative is discussed and agreed. Eviscerating and impoverishing local governments in 2021, just as the UK government seeks to leap forward and lead at COP26 in Glasgow, is wrongheaded, and it will not deliver.

Local governments across the world are increasingly demonstrating they want to accelerate their communities' net-zero transition. They must be assisted in this task, not undermined. In the end, it is the local communities that must plan the details of how their regions, cities, and towns respond to and plan decarbonization. It is the local dialogues and politics that translate national directives into local ownership and engagement, and this is where national plans become reality. It is often in our cities and towns that dynamic political activism can lead us towards a positive refashioning and reimagining of the net-zero future.

The proliferation of climate emergency declarations by regions, cities, and towns in the last two-plus years is a signal of this rapid policy shift and implementation of climate change policies in a raft of areas where local governments control the policies, determine the regulations, set expectations, and plan for a liveable, sustainable, resilient, inclusive, equitable future. National governments will still set the overarching goals and glidepaths. It is the local governments that will deliver new, greener services daily.

Cities as hubs of green innovation and sustainable liveability

Cities are where talent clusters, innovations are sparked, and cultures mix and evolve, and they are driving the climate change conversation and response. But cities are also the source of a great deal of carbon emissions. In the UK, for instance, the 63 largest cities and towns produce fully 50 percent of GHG output, with London alone producing 11 percent of the total. Cities that drive our economies today must also power the net-zero transformation. In the UK, as elsewhere, some do better and some worse. Ipswich is a leader at 3 tons of carbon per person per year, while for Londoners it is 3.6 tons, with steel-producing towns such as Redcar producing as much as 22.4 tons per person per year.

It is indisputable that cities are an essential component in the net-zero transformation. Overall, cities are more carbon efficient and are decarbonizing faster than suburbs or exurbs (BBC, 2019a). With their lead and commitment from communities, supported by governments, and new incentives, and short-, medium-, and long-term plans, we are more likely to reach net zero before we hit climate tipping points.

Most of the world's cities are in 2021 still failing to grapple with the net-zero challenge. However, increasing numbers of cities are sustainably lighting the way forward. The following sections provide illustrative examples of cities that have embraced a new green policy narrative and are moving towards achieving net-zero goals, while simultaneously making their streets more liveable, their environs more beautiful, and their locations more desirable for investors and citizens. Cities are helping set the agenda and shift the dialogue in scores of small and large areas of our daily lives, from building standards to transportation and mobility, to electricity supply and more.

Cities and the built environment

Cities and localities set building retrofit requirements, establish net-zero codes for new buildings, and require LED (light-emitting diode) lighting in buildings. All such steps bring net-zero goals closer to fruition. Cities can start with retrofitting their own building stock and require new builds to apply new codes and retrofits to apply new standards based on a progressive timeline. Many are already taking these steps.

For example, as far back as 2010, Frankfurt, Germany revised codes to require all city-owned buildings to meet the Passive House (Passivhaus) standard, a rigorous metric of building-energy efficiency. The commitment built upon Frankfurt's earlier 2007 requirement that all new buildings meet Passive House standards. Frankfurt is in the lead after 15 years of progressive climate change planning and codes and is demonstrating that shifting building design does not mean a downgrade but an upgrade, energy savings, efficiency gains, and cuts in GHG emissions.

Or take the case of Santa Monica, California, which approved the world's first ordinance requiring all new single-family homes be built to achieve net-zero energy by 2017. The city's action is more aggressive than California's Strategic Plan, which requires all new residential and commercial construction to achieve net-zero energy by 2020 and 2030, respectively.

In both Frankfurt and Santa Monica, we see city governments leading, far ahead of their state and federal counterparts.

Transport, mobility, and our cityscapes

Cities are setting standards for fleet electrification, EV charging stations, freight reductions, public transport changes, and many related steps that transform our built walking and biking environments.

Take Seattle, in the US state of Washington, which adopted a 'clean and green' fleet initiative way back in 2003 (Seattle.gov, 2019). Seattle now has one of the largest green fleets in the US. The green fleet plans have two goals: (1) to reduce GHG emissions by 50 percent by 2025, and (2) to use only fossil-fuel-free fuel by 2030. As part of this plan, the city is increasing the EV charging infrastructure. The cost to the city is an estimated and manageable US$28 million over seven years (Seattle. gov, 2019: 4).

Edinburgh, Scotland, my hometown, is also showing what is possible with positive planning processes and engagement. In 2019, Edinburgh released its ten-year masterplan, which includes tramlines, closure of bridges to car traffic, and a new cyclist- and pedestrian-only bridge to connect the Old Town and the New Town. The city is committed to net zero by 2030 and is replacing city vehicles with EVs; increasing charging stations; retrofitting buildings for efficiency and requiring Passive House standards for new builds; increasing green infrastructure; and reviewing all service agreements to bring them in line with the net-zero 2030 target and action plan (Edinburgh, 2020).

Other cities are restricting access and penalizing polluting vehicles to shift incentives. For example, Dublin, Ireland has banned freight vehicles with five or more axles from the city centre between 7:00 and 19:00 without a permit. Dublin's action has reduced freight vehicle traffic near the city centre by 80 to 94 percent, removing an average 4,000+ heavy goods vehicles from Dublin's roads (*Irish Examiner*, 2007). Paris, France has also begun the shift, announcing that it will completely phase out diesel cars by 2024 (Phys.org, 2017), citing pollution and public health concerns. The city plans to phase out all internal combustion vehicles by 2030. We can see these locations preparing to become post-car cities.

Powering our communities

Local governments can electrify the transition to net zero through LED smart streetlight programmes, municipal solar installations, renewable supplies, and incentives. Cities across the world – from Portland, Oregon to Malmo, Sweden to Hokkaido, Japan to Canberra, Australia –are committing to 100 percent renewable energy. They are setting deadlines for milestones and achievements and for implementing action plans.

Lighting our way forward can be literal and local. For example, in 2016, Belo Horizonte, one of the largest cities in Brazil, agreed to upgrade its 175,000 streetlights with efficient LEDs. To accomplish this task, a public–private contract was signed worth US$300 million over 20 years, one of the largest streetlighting contracts in Latin America. The project has resulted in 50 percent cost savings to the city and its residents, and significant GHG emissions reductions (Signify, 2019). San Diego, California offers another example of best practice. In 2017, the city announced a smart-city project which included the installation of 14,000 smart LEDs. The LEDs save the city an estimated US$2.4 million annually on energy costs, while a sensor network yields many additional benefits (Sandiego.gov, 2020).

Local governments can also promote renewable supplies by supporting communities seeking to harness net-zero power sources and so spur private investment for the transition. Harlaw Hydro outside Edinburgh, Scotland is an excellent example of community action and effect. The cooperative was established in 2012 to build a small hydropower station and generate electricity for sale to the grid, while throwing off up to 5 percent per year in income for local backers of green energy and supporting community projects in the village of Ballerno. The plan secured wide support and raised £403,000 through a share offering, primarily bought by villagers, that funded construction of a community hydro scheme at the Harlaw Reservoir. As of 2020, the power plant has been generating for five years and has generated 430 kilowatt-hours to feed into the grid, providing cooperative members with income at a rate far above bank deposits, while supporting local charity works and cutting GHG emissions.

Local governments are often well placed to understand the needs and pathways to net zero. Communities, properly consulted and represented, are more likely to be responsive to local plans than international and national directives alone. Amsterdam's good practices also show us a way forward.

In 2019, natural gas heated 90 percent of Amsterdam homes, producing a third of the city's GHG emissions. Amsterdam is committed to net zero by 2050. To achieve that goal, new developments are being built without natural gas infrastructure.

Existing neighbourhoods are having natural gas lines removed. Thus, 10,000 public housing units had their gas supplies removed in 2017 alone. Incentives are provided to homeowners to offset the cost of removing natural gas. By 2020, more than 100,000 homes had transitioned from natural gas (Amsterdam.nl, 2020). This difficult transition is possible because most of the city's inhabitants support the shift to electricity. If Amsterdam shows what is possible, so does Utrecht.

A better, localized, green, sustainable, liveable future

Utrecht offers an example of how the totality of a net-zero mindset and a redesign of the city centre can transform our cityscape, our experiences in it, and GHG emissions. The Dutch know how to make cities liveable, walkable, bikeable, and sustainable. Utrecht's transformation demonstrates this. In the 1970s, the old town was encircled by an ancient canal, which was filled in to make a 12-lane motorway. This car-centric plan split the city and was quickly and widely regarded as an historic mistake. In 2020, the city reversed this error and returned the Catharijnesingel canal to its original watery state. This move was part of a masterplan for the city that chose water and green space over cars. The plan includes the world's largest bicycle park at the city railway station, capable of storing 12,000 bikes. It also includes both the planting of green roofs in the city centre and its pedestrianization.

Utrecht is showing the way towards a better, more beautiful, sustainable, liveable city landscape. This shifting cityscape is underpinned by an energy plan, agreed in 2015, with all new homes in the city being energy neutral – ahead of the national goal. The city plan favours EV, public transport, and bicycles. Solar PV is used on all suitable roofs, and the city already has the highest rate of PV installations in the Netherlands (IRIS, 2020). Utrecht shows what is possible in our cities when we plan forward, work together, and shift our narratives. So, too, do many other cities I have no space to discuss here, such as Lexington, Massachusetts; San Francisco, California; Rio de Janeiro, Brazil; and Wuhan and Jihua, in China.

Nor do I have the space here to highlight all the successful local initiatives underway, but these examples show that localities and their leaders can do what it takes to reach towards net zero via tangible steps. These locally rooted, devolved, net-zero actions, and our engagement with them as citizens and voters undergoing our own narrative shifts, make what at first appears to be impossible, possible.

Constructing new tales and travelling with them into the future

Rethinking the tales we tell one another and the narratives we weave about climate change is an ongoing process. Some communities are demonstrating how to make it work. Discussions need to be grounded on facts, and the facts, once understood, can lead to dialogue and decisions on what should be done, who should take the lead in any area, and how the burden should be shared.

Evidence indicates this reality-based dialogue is possible, positive, inclusive, and productive. Such collective conversations and narrative reimagining can begin to partially replace the isolated, distorted, corrosive, AI-driven dynamics of web-based misinformation. The former is living and moving forward. The latter is wasting time and GHG emissions.

Different communities and groups of people (young and old) are replacing delusion with understanding and engagement. These conversations can be organized and facilitated, or spontaneous and dispersed. The steps being taken by localities and communities across the world, informed and supported by a better understanding of the climate change facts and transition glidepaths, suggest that narrative change is underway.

It remains to be seen whether this narrative leap can be made by all our communities, but it does appear as if the consensus is shifting towards scientific, reality-based, emotional, and ethical judgements on the glidepath ahead. We need to talk about it more, and then we need to act and demand action from our governments and communities, and our leaders and neighbours, economies and businesses. We all have a part to play in the new green narrative that is unfolding.

References

Amsterdam.nl. (2020) 'Policy: Phasing out natural gas' [Online]. Available at: www.amsterdam. nl/en/policy/sustainability/policy-phasing-out/ (accessed: 3 November 2020).

Anderson, J. and Rainie, L. (2017) 'The future of truth and misinformation', 19 October [Online]. Available at: www.pewresearch.org/internet/2017/10/19/the-future-of-truth-and-misinformation-online (accessed: 15 February 2021).

Axios. (2020) 'Poll: One-third of Americans are open to QAnon conspiracy theories', 21 October [Online]. Available at: www.axios.com/poll-qanon-americans-belief-growing-2a2d2a55-38a7-4b2a-a1b6-2685a956feef.html (accessed: 21 January 2021).

BBC. (2019a) 'Are cities bad for the environment?', 16 December [Online]. Available at: www.bbc.com/news/science-environment-49639003 (accessed: 16 September 2020).

———. (2019b) 'Scotland must 'walk the talk' on climate change', 17 December [Online]. Available at: www.bbc.com/news/uk-scotland-50808243 (accessed: 8 November 2020).

———. (2020) 'Kamala Harris asks Amy Coney Barrett: "Do you believe in climate change?"' 15 October [Online]. Available at: www.youtube.com/watch?v=TTNKg1jygpQ (accessed: 25 May 2021).

———. (2021) 'Climate change: Biggest global poll supports "global emergency"', 27 January [Online] Available at: www.bbc.com/news/science-environment-55802902 (accessed: 24 May 2021).

Bloomberg. (2020) 'Like a virus, QAnon Spreads From the US to Germany', 22 September [Online]. Available at: www.bloomberg.com/opinion/articles/2020-09-22/like-a-virus-qanon-spreads-from-the-u-s-to-europe-germany (accessed: 31 January 2021).

Bushnell, S., Workman, M., and Colley, T. (2016) 'Towards a unifying narrative for climate change'. Grantham Institute Briefing Paper No. 18, Imperial College London, April.

Business Insider. (2019) 'Conspiracy theories pose a new domestic terrorism threat, according to a secret FBI document', 1 August [Online]. Available at: www.businessinsider.com/fbi-document-conspiracy-theories-domestic-terrorism-threat-2019-8 (accessed: 30 January 2021).

Business Insider. (2020) 'The UK media regulator says a "Greta Thunberg effect" means more children are engaging in online activism', 4 February [Online]. Available at: www. businessinsider.com/greta-thunberg-effect-uk-children-online-activism-spikes-2020-2?r=US&IR=T (accessed: 24 January 2021).

Carney, M. (2020) Reith Lectures [Online]. Available at: www.bbc.co.uk/programmes/b00729d9 (accessed: 30 January 2021).

City of Edinburgh. (2019) 'Road map published for a net zero carbon Edinburgh by 2030', 22 October [Online]. Available at: www.edinburgh.gov.uk/news/article/12726/road-map-published-for-a-net-zero-carbon-edinburgh-by-2030 (accessed 15 September 2020).

Climate Generation. (2020) 'What is a climate story?' [Online]. Available at: www.climategen. org/take-action/act-climate-change/climate-stories/ (accessed: 30 October 2020).

Conway. K.A. (2017) Available at: YouTube: www.nbcnews.com/meet-the-press/video/conway-press-secretary-gave-alternative-facts-860142147643 (accessed: 15 January 2021).

Edinburgh. (2020) 'Edinburgh publishes 10 year draft city Mobility Plan', 10 January [Online]. Available at: www.intelligenttransport.com/transport-news/94582/edinburgh-publishes-draft-10-year-city-mobility-plan/#:~:text=The%20City%20Mobility%20Plan%20offers,quality%20of%20life%20for%20everyone (accessed: 30 January 2021).

Extinction Rebellion. (2020) [Online]. Available at: https://rebellion.global.

Francis. (2015) 'Laudato Si', 24 May [Online]. Available at: www.vatican.va/content/francesco/en/encyclicals/documents/papa-francesco_20150524_enciclica-laudato-si.html (accessed: 2 November 2020).

G30 (Group of Thirty). (2018). *Fixing the Pensions Crisis Ensuring Lifetime Financial Security.* Washington, DC: Group of Thirty.

Harvey, F. (2020) 'Young people resume global climate strikes calling for urgent action', *The Guardian*, 25 September [Online]. Available at: www.theguardian.com/environment/2020/sep/25/young-people-resume-global-climate-strikes-calling-urgent-action-greta-thunberg (accessed: 2 November 2020).

IFG (Institute for Government). (2007) 'Pensions reform: The Pensions Commission (2002–6)' [Online]. Available at: www.instituteforgovernment.org.uk/sites/default/files/pension_reform.pdf (accessed: 30 October 2020).

IRIS Smart Cities. (2020) Utrecht, the Netherlands [Online]. Available at: www.irissmartcities. eu/content/utrecht-netherlands (accessed: 15 September 2020).

Irish Examiner. (2007) 'New scheme to remove large lorries from city centre', 30 January [Online]. Available at: www.irishexaminer.com/news/arid-30295718.html (accessed: 3 November 2020).

Kaplan, S. (2020) 'The climate crisis spawned a generation of young activists. Now they're voters', *Washington Post*, 30 October [Online]. Available at: www.washingtonpost. com/climate-environment/2020/10/30/young-voters-climate-change (accessed: 2 November 2020).

Luntz, F. (2002). 'The environment: A cleaner, safer, healthier America', SourceWatch [Online]. Available at: www.sourcewatch.org/images/4/45/LuntzResearch.Memo.pdf (accessed: 24 May 2021).

NEA (National Education Association). (2019) 'The "Greta effect" On student activism and climate change', 19 September [Online]. Available at: www.nea.org/advocating-for-change/new-from-nea/greta-effect-student-activism-and-climate-change (accessed: 2 November 2020).

New York Times. (2020) 'Welcome to the rabbit hole', 16 April [Online]. Available at: www.nytimes.com/2020/04/16/technology/rabbit-hole-podcast-kevin-roose.html (accessed: 30 October 2020).

Pew Research Center. (2017) 'The future of truth and misinformation online', 19 October [Online]. Available at: www.pewresearch.org/internet/2017/10/19/the-future-of-truth-and-misinformation-online (accessed: 30 October 2020)

Phys.org. (2017) 'Paris wants to phase out diesel cars by 2024', 12 October [Online]. Available at: https://phys.org/news/2017-10-paris-phase-diesel-cars.html (accessed: 3 November 2020).

Reisman, K.C. (2008) *Narrative Methods for the Human Sciences*. Thousand Oaks, California: Sage Publications.

Sandiego.gov. (2020) 'Smart streetlights program' [Online]. Available at: www.sandiego.gov/sustainability/energy-and-water-efficiency/programs-projects/smart-city (accessed: 3 November 2020).

Scottish Government. (2020) 'Big Climate Conversation: report of findings', 30 January [Online]. Available at: www.gov.scot/publications/report-findings-big-climate-conversation (accessed: January 2021).

Seattle.gov. (2019). 'Green fleet action plan: An updated action plan for the city of Seattle' [Online]. Available at: www.seattle.gov/Documents/Departments/FAS/Fleet Management/2019-Green-Fleet-Action-Plan.pdf (accessed: 3 November 2020).

Segal, M. (2019) 'To fix the climate, tell better stories'. *Nautilus*, 15 August [Online]. Available at: http://nautil.us/issue/75/story/to-fix-the-climate-tell-better-stories-rp (accessed: 30 October 2020).

Shiller, R. (2019) *Narrative Economics*. Princeton: Princeton University Press.

———. (2000) *Irrational Exuberance*. Princeton: Princeton University Press.

Signify. (2019) 'Belo Horizonte realizes 50% electricity cost savings by upgrading more than 182,000 streetlights to LEDs from Signify', 15 August [Online]. Available at: www.signify.com/global/our-company/news/press-releases/2019/20190815-belo-horizonte-realizes-electricity-cost-savings-by-upgrading-to-led-streetlights-signify (accessed: 3 November 2020).

Taylor, M. (2020) 'Extinction Rebellion: How successful were the latest protests?', *The Guardian*, 11 September [Online]. Available at: www.theguardian.com/environment/2020/sep/11/extinction-rebellion-how-successful-were-the-latest-protests (accessed: 2 November 2020).

TechCrunch. (2019) 'Bad news: Facebook leads in news consumption among social feeds, but most don't trust it, says Pew', 2 October [Online]. Available at: https://techcrunch.com/2019/10/02/bad-news-social-media (accessed: 30 October 2020).

Thunberg, G. (2019a) Speech at World Economic Forum, 25 January [Online]. Available at: www.youtube.com/watch?v=RjsLm5PCdVQ&t=2m23s (accessed: 2 November 2020).

———. (2019b) Speech at the UN Climate Action Summit, 23 September [Online]. Available at: www.npr.org/2019/09/23/763452863/transcript-greta-thunbergs-speech-at-the-u-n-climate-action-summit (accessed: 2 November 2020).

Veland S., Scoville-Simonds, M., Gram-Hanssen, I., Schorre, A. K., El Khoury, A., Nordbø, M.J., Lynch, A.H., Hochachka, G., and Bjørkan, M. (2018) 'Narrative matters for sustainability: The transformative role of storytelling in realizing 1.5°C futures'. *Current Opinion in Environmental Sustainability*, 31: 41–47 [Online]. Available at: https://doi.org/10.1016/j.cosust.2017.12.005 (accessed: 30 January 2021).

Vice. (2020) 'QAnon presence grows at UK anti-lockdown protests', 19 October [Online]. Available at: www.vice.com/en/article/7k9zex/anti-lockdown-march-uk-qanon-conspiracy (accessed: 30 January 2021).

Yale. (2020) 'Yale climate opinion maps 2020', 2 September. Available at: https://climatecommunication.yale.edu/visualizations-data/ycom-us (accessed: 24 January 2021).

8

ON DECARBONIZATION, ECONOMIC GROWTH, AND A JUST TRANSITION

I believe that the pandemic has presented such an existential crisis – such a stark reminder of our frailty – that it has driven us to confront the global threat of climate change more forcefully and to consider how, like the pandemic, it will alter our lives.

Fink, 2021

It is now possible to decarbonize economic growth and to achieve deep reductions in greenhouse gas emissions while increasing economic activity and prosperity.

Jotzo, 2016

There is a moral responsibility to ensure a more equitable distribution of the benefits and risks associated with the choices that we make. The notion of just transitions suggests that these diverse changes should be managed so as to bring about environmental sustainability, social equity, and economic development in ways that 'leave no-one behind'.

Schwartz, 2020

Creating a decarbonized economy requires us to recognize the crisis and to make huge leaps in how our societies and economies operate (Fink, 2021). The contours of global growth and globalization as we decarbonize and renewably energize the global economy should not be the same as the existing political economy model. The across-the-board, multidecade investment cycle needed to ensure we limit the global temperature rise, as a far as is possible, to 1.5 degrees or well below 2 degrees Celsius, can instead result in a Green Globalization 2.0. The transition has already begun in some sectors, such as energy and renewables, and with vehicle electrification. It must spread rapidly across all industries, economies, societies, and

DOI: 10.4324/9781003037088-9

communities. The transformation can be a just transition (Schwartz, 2020) and, if it is well constructed, can help ensure broader-based economic growth (Jotzo, 2016).

As I have stressed, green industrial policies, carbon pricing, and re-regulation are needed to speed the rate of diffusion and steepen the S-curve in many sectors. Too often, in 2021, the take-up of existing technologies that we know can bend the curve of GHG emissions is too slow (such as in agriculture or construction). This must be addressed nationally, regionally, and locally. The current technological evolution and diffusion exhibit a faster adoption rate than in past revolutionary technological shifts. While it took over 100 years from invention for the internal combustion engine to be adopted widely in the world, the take-up and use of the internet rocketed much faster, with advanced economies adopting it from a standing start within 20 years (Andres *et al.*, 2007). In 2021, there is good reason to suppose that the rate of diffusion and adoption of technologies could be similarly fast in many sectors and countries, given that technology transfer, investment, and skill building occur with the support of advanced economies.

As diffusion rates steepen, costs fall, adoption surges, and innovation is sustained, the contours of Green Globalization 2.0 and our ability to secure targets will be positively impacted by increased economic growth rates caused by the transition, as multiple national and regional GNDs power the shift in our economies and societies. However, the notion that growth can accelerate and power the green transformation is disputed.

Growth or no growth

One camp – in which I stand most of the time – asserts that it is possible to have sustained, faster economic growth, while progressively decoupling it from GHG emissions. Continued green growth is, according to this position, necessary to fund the net-zero transition, address inequality, increase resource transfers, and tackle fairness. This Green Globalization 2.0 stance posits that it is only with a growing economy that extremely difficult political economy trade-offs can be financed and can underpin sustainability. Another camp proclaims the reverse – the need for degrowth – for some parts of the global economy to slow to a standstill and reverse, to secure sustainability and planetary survival. Advocates of degrowth view economic growth, as currently practised, as an unrestrained use of finite resources for private profit with marginal concern for the GHG and planetary outcomes, and thus as incompatible with humanity's survival. Degrowth is needed, it is asserted, to readjust the contours of globalization and remake the economy.

Optimism or doom and disaster – a false choice

The two positions – Green Globalization 2.0 and growth versus degrowth – pitch human and technological optimism against planetary and political pessimism. In this chapter, I place myself in the former with a foot in the latter. I believe in the possibilities of technological breakthrough, innovation, human ingenuity, and our ability to act and correct our environmental failures. I believe we can have green economic growth, decarbonization and economic decoupling, and a new Green Globalization 2.0. The contours of that new globalization should be different, markedly better, sustainable, resilient, robust, ethical, and inclusive. I also recognize that we cannot act as if the resources of the planet are limitless. They are not. I stand with those who seek to address issues of inequality, fairness, and a secure and just transition. The green future will not come about by accident or due to the munificence of millionaires and billionaires. Unsupervised, markets and firms will not deliver the desired outcome. Rather, it must be designed to make it just for all.

New contours for a green tomorrow

We need to consider new contours for globalization, environmental sustainability, equity, and the form of economic progress we collectively construct. We should avoid replacing rapacious neoliberal globalization with sustainable electronic servitude. The future must be better for humanity than just being isolated before a screen and being serviced by vast behemoths such as Amazon, Facebook, and Google. These giants are not our friends. They are our tribal enablers, our personalized dark rabbit holes. As we work to secure a net-zero future, Amazon and Facebook should not be left unregulated and unsupervised and allowed to become more dominant.

It is possible in the decades ahead to construct a Green Globalization 2.0 that harnesses and redirects economic markets to achieve the common goals of decarbonization. Tim Jackson, the leading degrowth advocate, is correct that unregulated, polluting, inequitable, non-green growth coupled with distorted (too-low) wage rates and massive increases in inequality have blighted our societies and damaged our politics and cohesion while undermining economic and ecological sustainability and warming the planet.

A reimagined globalization must be constructed on green growth that is net zero or carbon negative. Renewables will power green globalization and are already doing so, as we have seen, and utilities are making the leap as well. The utilities sector, further along the journey than others, illuminates the way forward to a more sustainable future. The construction of Green Globalization 2.0 will require greener cement (see Box 6.9, page X), with green, energy-efficient buildings that cost less to maintain and heat and which are better places to live and prosper. It is not that we will stop building. To the contrary, we must replace or retrofit our global urban, suburban, and rural environments and living spaces.

Feeding the planet requires huge changes in agriculture and land use. The food and sustenance we require to live a greener, globalized, sustainable life must be produced. This must be done in a manner that is markedly different from today's destructive monocultures and GHG-belching systems and animal choices. Here again, in agriculture many of the solutions are within our grasp. We can shift practices, eat less meat, farm more sustainably, shift incentives, announce phase-outs, innovate, and redesign. A reseeded Green Globalization 2.0 will perhaps not be as frenetic; we might travel somewhat less and to different locations, or shift our plans and consumption patterns. This is not the end of globalization; rather, it is a reimagined, localized, rerouted globalization in which we pay the real costs of our decisions and choices.

The redesigned green growth and GND that will help get us there, in the US, Europe, and elsewhere, will be grounded upon refocused industrial policies that can electrify sustainable growth and prosperity. The required massive, multidecade investment flows into Green Globalization 2.0 can result in economic multipliers and address Lawrence Summers' (2014) secular stagnation concern. Green Globalization 2.0 can address secular stagnation through green investment, reskilling, better pay, a larger workforce of motivated workers, increases in productivity, increases in the rate of economic growth, and the maximum potential growth rates in many countries, without inflationary pressure.

Green Globalization 2.0 will be supported by the world's central bank monetary policies. Extremely low (or negative real) interest rates are a time to redesign and to invest in green growth opportunities and a sustainable economic future. What this is called matters less than the reorienting of policies to embed net-zero goals.

Green Globalization 2.0 can help address the populist anger in many countries in 2021. That anger is understandable and has been driven by stagnant wages and poorer tomorrows, destructive local outcomes of globalization that have left too many behind (Rodrik, 2017), diminished or destroyed the middle classes, and fractured our societies into 'us' versus 'them' and caused tribal-cultural splits (Collier, 2018). The contours of net-zero Green Globalization 2.0 and of our national economies can be more equitable and more supportive of societal cohesion and progress. Investment in GNDs should be matched by a reappraisal of burdens and who carries them within our societies. These adjustments, essentially an assurance of a more equitable transition, can be bearable. If we understand and agree on the climate change facts and alter our narratives so we can move towards the green goal, we can shift our conversations with one another and act collectively and dynamically and reimagine a dynamic, prosperous net-zero future.

Constructing Green Globalization 2.0

Green Globalization 2.0 is capitalism yoked to the green transformation and forced to live within a carbon budget and real planetary boundaries. This form of green globalization, as Mazzucato (2019) has observed, 'should be about co-creating, co-sharing markets alongside the private sector'. Steering investment towards public net-zero missions shared by a redirected, energized private sector can stimulate investment and innovation. This can be done without micromanaging. Governments will set a direction and then use the full array of government policy instruments to fuel bottom-up experimentation and exploration by public scientists and private investors. This has worked in the past and can do so again.

This chapter looks at the contours of green, reseeded globalization. It argues that it is possible to speed growth in some sectors while activities in others shrink. Green growth can be more inclusive and broadly based than the 'winner takes most' app and IT growth in which a few have become extremely wealthy and the rest have benefitted far less, if at all. Green Globalization 2.0 can provide a solution to secular stagnation and economic distress. This reseeding will be the basis for our future prosperity.

Simultaneous with GNDs, governments should revisit issues of burdens and fairness to ensure that opportunities to game the system are minimized. Governments should ensure we all play our fair part in shouldering the cost of the transformation. When the burdens are seen to be visibly fair, people will work together and support the war against carbon. Conversely, if the burdens remain unbalanced, our net-zero stories and narratives cannot be easily constructed or agreed, and we will still be split and unable to act with dispatch.

I believe that fairness must play a greater role in our policy calculus, replacing erroneous *homo economicus*, the empathy-lacking utilitarian robot, with *homo economicus sympatico*, an avatar that operates informed by economic choices and incentives but also by an ethical and moral compass in making decisions.

Green New Deals for all of us

In 2020, many governments signalled they were commencing national GNDs to alter the nature and wiring of globalization going forward. Some of the GNDs go further faster than others.

We have already seen that green growth is possible and occurring. GNDs can provide an industrial policy energizer to the process. For much too long the standard refrain by critics of greening industrial policy and of acting on climate change has been that there is, as Levitz (2020) notes, 'a hard trade-off between tackling the crisis of economic stagnation and that of ecological decline'. The critics are wrong. To the contrary, the decarbonizing, re-energizing, revitalizing effects of multiple national GNDs, and regional and coordinated global policies on climate change will drive economic growth, not hinder it. GNDs will accelerate investment, not lower it, and increase innovation, not stymie it.

All GNDs are about funding the innovative green tomorrow supported by both public institutions and private sector investors and actors.

Governments can afford to invest in green reseeding

In a world of low interest rates, which appear to be here for the foreseeable future, there is little danger of a runaway debt spiral due to debt repayments. Governments can therefore invest in the construction of Green Globalization 2.0 using ultra-long-duration government bonds and invest in the green infrastructure (power, distribution, and so on) reconstruction and redesign needed to spur decarbonized globalization. As we have seen, government policy sets the rules, guardrails, and glidepaths, and this has already begun to unleash private sector decision-making processes and investment dynamics. Governments should invest now to secure net zero by 2050 and support our collective planetary goals.

Governments can establish recharging networks or incentivize their creation; they can support green innovation, the creation of a hydrogen infrastructure, and agricultural and industrial shifts in economy-wide practices and towards net-zero goals. Some of this is regulatory action and has no direct cost to governments. Some has costs attached but is needed, nonetheless. In this very low and negative interest rate environment, green investments are sensible and will pay off. Green investments will pull forward the rate of transition and trigger private investment flows across the economy, creating multiplier effects and feedback loops within the economy.

The GNDs will speed Green Globalization 2.0 and support the rerouting of markets away from polluting goals towards net-zero goals. This has begun and should be further energized at COP26 and via national and collective decisions on GND industrial policies and net-zero goals and implementation.

Policymakers, from President Biden to President Macron to President Xi, understand this urgency and are linking the redesign of globalization to the green tomorrow and transition. The question is not: Can we green our growth? It is: How fast can we do it? Leaders are focusing on a reseeding and regrowth of green market technologies and innovations and using positive narratives that convey newly possible avenues that can help us reimagine our better futures together. This reseeding can re-energize productivity and economic growth. We are not at the end of innovation and growth but, rather, at the start a new phase of innovation and dynamism.

Productivity has been languishing for years, as Robert Gordon (2016) has shown. However, Gordon's overall case – that we are at the end of innovation and have exhausted our engines of growth – does not convince. It is too pessimistic and fails to account for the dynamic effects of crises once recognized, of populations once engaged, and of governments once politically committed and investing in the regreening processes.

A greener, more productive economy

Part of the solution to Gordon's productivity paradox and Summers' secular stagnation conundrum in the US, the UK, and elsewhere is to harness more equitable green growth dynamics. The transition to net zero can begin to bridge wage gaps, speed productivity, and provide a fair living wage to workers, and in doing so speed up economic growth and increase maximum potential growth.

For too long, wages have been stagnant and costs rising, while the gains of productivity have been captured by fewer and fewer. As a result, workers have been demotivated and disincentivized. I would be. Wouldn't you? (Akerlof and Yellen,

1990). With many economies operating below full potential for decades prior to Covid-19, and now because of the pandemic, we need to invest, restructure the economy, and allow creative destruction and refashioning.

Governments, bond-funded, negative-interest-rate financed, and central-bank supported, should seize the opportunity to invest at very low cost or negative rates. Austria, for instance, has sold a 70-year bond that pays negative rates. In such an environment, failing to borrow and invest is irresponsible. Take on the debt and invest in Green Globalization 2.0.

Critics worry the US, the EU, and Asian GNDs might cause our economies to eventually run hot for a while. But that's not likely, and modest inflation would not be unmanageable if it occurs. Faster wage growth could, rather, help workers regain the ground they have lost in the last 50 years. Better-distributed green wage growth, fuelled by green investments, resource transfers, and rapid diffusion supported by governments, should be pursued and welcomed, not worried over. These investments, in turn, can help finance yet more investments in a cycle of renewal and reimagining. Properly overseen and supported, the net-zero transition will also help engender higher-productivity work through a reskilling of workers and new professions and green specialties.

But what of the cost of GNDs?

The cost of GNDs may appear high in the short term, but they are manageable (at negative rates) and they will pay planetary and economic dividends. Of course, wars are never cheap; they are often painful, difficult, and longer than expected. But such crises (and wars and pandemics) are also societal levellers (Scheidel, 2018). Crises open new possibilities and make the impossible possible. Wars, for example, are destructive and dynamic; they create new coalitions, speed the rate of change, see societies pull together, and open up possible new futures. In 2021, we have no option but to respond to the climate crisis, either now, in a measured and managed process, or later, in a disorderly one, as the central bankers warn us.

A response to debt and deficit hawks

We should respond when economists we know, and I know far too many, appear fixated on the current short-term account deficit in the next quarter or next year, and ask: 'Can we afford it?' We should respond: 'Compared to what?' Is it worth it compared to societal and ecological collapse? Is it worth it to help arrest the mass extinction of a third of nonhuman species on the planet? Is it worth it to avoid the dieback of the Amazon rainforest, the death of boreal forests, and the collapse of the Greenland ice shelf?

Looked at from this perspective, funding GNDs is an entirely manageable cost when placed against these risks. When that funding is available at near-zero interest rates or negative real rates, we can afford to invest for the future. Not to do so would be an abrogation of our duty to the planet and our children's future. GNDs are not simply a cost; they are a bargain insurance policy for humanity. They are an essential, effective, productive investment policy amplifier and accelerator.

GNDs as policy accelerators

GNDs will act as green reseeding and regrowth policy accelerators. They are designed to realign, and embed government policies with the collective net-zero goals to hasten their achievement. GNDs are already triggering public policy and market responses and reactions, pulling forward private investment flows. By committing large amounts of public resources, governments are signalling to the economy and markets. GNDs are changing the predictions of investors even before they directly affect the markets, and they are helping to speed the conversion, amplifying public policy effects and reach.

On the public side, the altered policy narrative also includes the explicit recognition and prioritizing of green considerations and policy alterations, for instance:

- By the creation of carbon councils and monitoring regimes
- By the EIB in its announcement of green investment goals
- By the European Commission's goals for the Covid-19 green rebuilding
- In the adjustment of scores of central banks' policies and how they interpret their mandates – via the NGFS process
- In the ECB, Netherlands Central Bank, or Bank of England shift on climate change.

The shift in process is also seen on the private sector side, in, for example:

- The Climate Action Network announcement on net zero
- The We Mean Business coalition in Europe[1]
- The Transform to Net Zero initiative in the US[2]
- The clear statements by Larry Fink, BlackRock chairman and CEO; the government of Singapore-owned investment company Temasek; and others
- The hundreds of firms committing to TCFD requirements and net-zero goals.

Thousands of organizations and firms are making the leap to net-zero planning. The number of corporate commitments has doubled in less than a year. By September 2020, companies worth over US$11.4 trillion dollars had committed to net-zero goals (UNFCC, 2020). We must monitor those commitments and make sure they turn into facts on the ground, but they are far from meaningless and signal that publicly driven and private market shifts are speeding up.

A great many such steps, and many more regulatory actions, are helping move market narratives and expectations, country to country and globally. The GNDs, regulatory and policy shifts, and public and private measures reinforce one another and create feedback loops, pulling forward expectations, planning, and market and investor judgements and actions.

Sustainable growth is possible and essential

This is not degrowth. Rather, it will be a different sort of decoupled, more balanced, decarbonized growth. This growth will require that some categories of economic activity grow massively, such as those associated with the production and distribution of clean energy, electrification, retrofitting, construction, green agricultural processes, and so on. Other, polluting sectors will shrink, and firms will close. This should happen simultaneously with the removal of subsidies and the rapid move away from fossil fuels, with polluting firms going bust and stranded assets being seen, if polluting firms do not begin to quickly transform, as some are doing, into clean energy utilities.

A more green-tinted, carefully invested, environmental, social, and governance-driven growth rate is already visible and will further accelerate the pace at which clean energy, clean transport, and other innovations replace fossil fuels, given the right regulation, oversight, and market signals. Higher economic growth rates will result in higher levels of investment channelled into green technologies and innovations, not less. This is already visible in some sectors, with the positive economic feedback loops boosting the rate of diffusion adoption and growth.

Renewed Green Globalization 2.0 as a partial answer to populism

Green growth may also help countries address societal stresses and populism as green growth, jobs, wages, and opportunities are spread more evenly across cohorts, unlike the tech and digital economy, where most benefits are captured by the few. If you believe that populism is a cultural phenomenon but also fundamentally an economic story of dislocation and disappointment, of low wages and higher costs, of fewer opportunities, of truncated lives, and of anger, then it should be addressed via strong, broad-based, more equitably shared green growth. US president Joe Biden sees the urgency of addressing both climate change and economic dislocation and anger. Dealing with economic dislocation by shifting government policies to align with climate change goals can help ameliorate the anger.

Biden's government-wide, green policy leap

The Biden administration's shift on climate change policy is part of a wholesale realignment of all government policies to include climate change goals. This realignment is visible in the repeated identification of climate change as a top policy priority for the administration by the president and by all top cabinet entities and officials, including the National Security Council, the secretary of the treasury, and the secretary of state. The importance of this dramatic change cannot be overestimated. The narrative at the very top of the US government has shifted. All levers can begin to be used to achieve the climate change goals.

The Biden administration's US GND signals the seriousness of the policy shift. At US$2 trillion, it is the most ambitious green industrial policy ever proposed by a successful US presidential candidate. If passed into law, it would trigger a staggering 20 times the investment of the Obama administration in green technologies and would swiftly alter the policy narrative and America's collective climate change story.

The Biden administration's US$2 trillion GND and investment policy will be accompanied by major regulatory and incentive shifts. Within hours of being sworn into office on January 20, 2021, Biden signed an executive order committing the US to re-joining the Paris Climate Agreement. The administration will also set aggressive net-zero GHG goals and take many other regulatory steps. The Biden proposal is radical by American standards. Why? Because of the proposed scale of green investments but also because 'the core of the Green New Deal, if you just look at the projects, is … one of the largest interventions in US industrial policy in a long time' (*Atlantic*, 2019a).

President Biden's embrace of an American green industrial policy dovetails with efforts in other countries. We see versions of it in the UK, China, Singapore, Europe, including France, and elsewhere. Countries increasingly understand the necessity of setting directional goals and adjusting incentives and penalties with respect to the climate and the environment, something which all governments do in other areas all the time.

As Cohen and Delong (2016) note, America has always sought to adjust direction and shift investment goals, as good public policy ought to as an economy evolves, even if markets remain dominant:

> Yes, there was an 'invisible hand'. … But the invisible hand was repeatedly lifted at the elbow by the government and re-placed in a new position from where it could go on to perform its magic.
>
> *p. 2*

US President Dwight D. Eisenhower, for instance, commenced a massive investment programme far larger than that undertaken by FDR, and it underpinned the prosperity of the 1960s. The Biden green transformation can do likewise for millennials and Gen Xers and set the stage for sustained green growth in the 2020s and 2030s.

President Biden's ability to enact his green transition policy might be hampered somewhat by a narrow Democratic margin in the US Senate, and thus major spending bills might pass only with difficulty or might even stall. This would be a real setback, a weakening of the economic multipliers of his policies. However, Senate obstinance and backwardness may not be fatal to the US green transition policy.

The Biden administration can still rapidly shift incentives, regulations, and penalties, via administrative action or executive order, without passing spending bills. A raft of greening regulations and actions is expected to significantly change investment assumptions made by private actors and investors about the risks and returns of green versus brown companies and sectors. In a whole host of areas, the greening of the economy and transition are set to accelerate and spur growth and pull forward investment decisions and plans.

The Biden green policy agenda is signalling a new beginning in America. Thus, in addition to rejoining the Paris Agreement:

- The US has announced a net-zero goal, and will establish the route and glidepath to 2050
- All cabinet officials have repeatedly stressed that climate change is a top priority across the administration
- President Biden has appointed former Secretary of State John Kerry as climate czar to lead US reengagement (Kerry signed the 2016 Paris Agreement on behalf of the US).

Further, the administration:

- Will begin to take steps on internal carbon pricing assumptions and mechanisms (a tax or ETS)
- Will change the discount rate being applied for government planning to no higher than 2 percent
- Is acting on Corporate Average Fuel Economy standards and EV support
- Is reregulating CO_2 by the US EPA
- Is re-establishing regulation of methane emissions
- Will add to fracking regulation
- Is banning new oil leases on federal land
- Is increasing construction standards
- Is addressing efficiency standards
- Is addressing incentives for solar and wind.

None of the above regulatory steps requires a huge commitment of resources at the outset, but all signal a meaningful shift. Taken individually they are small steps, but when viewed as a whole and seen as part of what amounts to a complete strategic realignment of US climate policy, they should be seen as what they are: a fundamental reorientation of the US policies towards climate change and a commitment to net-zero goals, timelines, and the green industrial transition.

Getting policies aligned triggers rapid corporate responses

Visible climate change policy shifts can trigger rapid public sector and corporate and market responses across countries, regions, and industrial sectors.

Major US businesses understand that policy lever incentives are going to shift dramatically in 2021. For instance, in a huge shift, the conservative and staunchly pro-business US Chamber of Commerce announced its support for carbon pricing and trading. This corporate shift will gather momentum in 2021 and beyond, lest firms and their CEOs get left behind. Many have already made the leap, shocking their rivals.

General Motors stuns the car industry

General Motors CEO Mary Barra was one of the first American leaders to respond to the change in the policy landscape in 2020, announcing, 'President-elect Biden recently said, "I believe that we can own the 21st century car market again by moving to electric vehicles." We at General Motors couldn't agree more' (*New York Times*, 2020a).

The firm has announced an increase in its commitments, including accelerating plans to introduce electric cars and trucks over the next five years and raising its EV investments to US$27 billion by 2025, up from a previous budget of US$20 billion. Ms Barra was unequivocal in her shift: 'Climate change is real, and we want to be part of the solution' (*New York Times*, 2020b). GM's explicit reorientation is only the first of many to come as those parts of corporate America that were waiting to see the outcome of the 2020 election shift strategies to align with the Biden administration green incentives and approach.

Shocking the global automotive industry, Barra announced in January 2021 that GM would stop building gasoline-powered cars by 2035. This is nothing short of an earthquake. The largest American car builder announcing such a step changes the entire US market. All other car manufacturers, except for Tesla, must now play catch-up or lose market share and shrink. GM's Barra is betting all on leading in the transition. She is not waiting and discounting the future. She can see it, seize it, manufacture it. That is what real leaders do.

A growing green economy

Importantly, despite the significant damage done by the Trump administration to US progress on climate change, the Biden administration is not building on ruins. This is because the corporate shift on climate change continued throughout the last four years, led by firms and actors choosing to bet on the future, not the past.

Just how big is the green economy already? Many US firms, markets, and investors are, as I have shown already, making the transition. In 2019, Georgson and Maslin (2019) estimated that the US green economy represented US$1.3 trillion in annual sales revenue and employed nearly 9.5 million workers; both figures have increased by 20 percent in only three years. The survey also found that the US has a greater proportion of the working-age population employed (4 percent) and higher sales revenue per capita in the green economy than other OECD states. Overall, Georgson and Maslin (2019) estimated that revenue in the global green economy was at least US$$7.87 trillion in the fiscal year 2015/16. They conclude that the green economy appears to be underestimated as a driver for growth and that many countries have a huge potential to generate higher green employment and growth from the transition. As they state, 'Given the climate change emergency and the employment slump in fossil fuel industry, it only makes sense that future investment should focus on growth in the green sector' (phys.org, 2019).

The impact and reverberation of the Biden GND; regulatory shifts; China's, Europe's, and Japan's net-zero goals; the building COP commitments and progress; the collective market responses; and the proportion of the economy that will be greening and greener will affect the rate policy and business changes. As I have observed, change can be slow, almost imperceptible, until suddenly a tipping point is reached, diffusion S-curves are steepened, and narrative leaps are made, markets shift, and investors move.

Importantly, in 2021, the contours of the green economy already visible do not appear to be aping the distortive and damaging lines of the existing inequitable, neoliberal paradigm. There is a unique opportunity to foster economic growth along new pathways that extend prosperity and reassure voters that tomorrow need not be worse than today but, rather, can be better, greener, and more liveable.

Green industries are growing faster than the overall economy. For every percentage-point increase in an industry's green intensity (the share of employment in green jobs), annual employment growth was higher. More green jobs are good news. Projections for the next ten years suggest continued jobs benefits from green intensity. States with greater green intensity also generally fare better in economic downturns. Those states that are engaged and committed to net zero are delivering better economic outcomes. This dovetails with what we should expect. When policy and strategies are aligned in the public and private sector towards net-zero goals, they reinforce each other and speed growth and resilience. In contrast, states with a polluting, old-industry mindset that do not support the transition and misspend scarce resources get poorer outcomes.

Many green jobs are also accessible to workers without a college degree. This dynamic is positive and welcome. We need to bridge the gap between the haves and have nots and to assuage voter anger with broad-based, sustainable economic growth. Green growth can extend across the economy and need not necessarily be concentrated in the pockets of the lucky and the few. The green seedlings of growth are being planted in America, and the signs are positive.

In the US, the Economic Policy Institute (EPI) estimates of the impact of the Biden GND suggest large positive employment effects. The EPI estimates that major investments in infrastructure, clean energy, and energy efficiency could support between 6.9 million and 12.9 million US jobs annually by 2024 (EPI, 2020). A US green job boom is already underway. In 2018, 335,000 people worked in the solar industry and more than 111,000 worked in the wind industry, compared to 211,000 working in coal mining or other fossil fuel extraction industries. Clean energy employment grew briskly by 3.6 percent in 2018, adding 110,000 net new jobs (4.2 percent of all jobs created in 2018) (Forbes, 2019).

As we reindustrialize, decarbonize, and redesign the net-zero future, it is not a future being constructed on TikTok or by influencers Instagramming or spreading misinformation on Facebook or 4chan. These are low-capital-intensive, winner-takes-all platforms of questionable productivity or social and economic utility.

Instead, the green industrial revolution is being built on, and will rely on, skilled manufacturing and on wind, solar, EV production, retrofitting, and redesign – real jobs for real people across all communities. The green-skilled manufacturing and service jobs are markedly different from the low-wage, low-skill service jobs that upset so many and leave them searching for escape.

I have said previously that economic crises, pandemics, and wars can be both destructive and creative and can be periods when old ways are abandoned and new models and approaches adopted, social contracts renegotiated, and economies revitalized. World War II provides further historical evidence that the US green transformation policy, and by extension the sum of national GNDs, can have dynamically positive effects that materially change the economy for the better. World War II expenditure shows why.

Targeted expenditure can lead to faster growth, more employment, and higher productivity

Instead of being a demand-and-supply dynamic, as one would see in a 'normal' economic recovery, the scale of massive structural redesign of the net-zero economy in the next decades will shift the potential growth rate of the economy upwards as a whole, rather than run an existing polluting, fossil-fuel-based economy too fast. The global net-zero process and GNDs will draw many more workers into productive employment, just as occurred in World War II.

Mason and Bossie (2020) show that during World War II public spending drew 13 million additional workers into employment, with no reduction in the size of the overall workforce. The war and massive investment caused people who had long given up looking for work to return to productive, well-paid employment. Many of these new employees were women (immortalized by Rosie the Riveter). Other workers moved from less productive, poorer paid jobs to better prospects in industrial production. Still others came off the sidelines and idleness.

One of the problems that has long bedevilled America and many other economies is the historic underemployment of the total workforce. For years, the US workforce participation rate has slipped and stayed worryingly low. Only Japan has been able to break this cycle through massive stimulus, pro-women policies, and cultural factors. Far too many US and European workers are underemployed, discouraged, or have stopped looking. Ludwig (2021) has shown the real unemployment rate, if calculated to include discouraged and underemployed workers and those paid less than a living wage, in the US is much higher than normally reported. If the real numbers are anything close to Ludwig's numbers, this is another explanation for the low potential growth rate and secular stagnation problem. It is intuitively not surprising that certain otherwise healthy groups will withdraw from the workforce if wages are persistently too low, costs (such as health, education, and childcare) are consistently too high, and good opportunities do not exist. In such an environment the potential growth rate will be lower and productivity will suffer. The green transformation and GNDs could partially address economic populist anger and be an answer to Summers' secular stagnation because the picture of employment growth and wage growth may be different and more broadly based, as the OECD finds:

> Green policies will reshape labour markets in ways that create new opportunities for workers, but also new risks. … A successful transition towards green growth can create new opportunities for workers, if the associated challenges are managed well. Jobs will be created in green sectors, and jobs will be destroyed in their 'brown' counterparts with high environmental footprints; the knock-on effects on employment in other sectors can also be significant.
>
> *OECD, 2017: 3*

In the early phases of the transition, the new jobs will be historically green jobs in sectors such as green agriculture, sustainable construction, sustainable forestry, public transport, renewable energy, recycling and waste management, clean industries and carbon capture, and in federal and local government activities.

Ultimately, and faster than we may think, the Green Globalization 2.0 of the future will stretch across all sectors and economic activities. All sectors will be impacted and will have to adjust to be secure and thrive or they will fail in the net-zero environment. The green transition is the motor of our future sustainable global growth. It is those green markets, jobs, higher wages, new technologies, dynamic firms, and innovations that will provide well-paid, skilled work. Green Globalization 2.0 can herald a sustainable, concrete, step-by-step, solar-and-wind powered electrified road to net zero. The transition can extend the current rapid industrial diffusion we have seen in the computer and digital revolution to our green revolution and transformation through and beyond 2050. Here, again, the GND net-zero transformation may reflect what occurred in World War II.

Not only did massive war expenditure in the 1940s drive a rise in well-paid employment, it also turbocharged productivity, innovation, and dynamism in the economy. When the war began, the total factor productivity of the US economy – that is, the economy's productivity growth – was less than 1 percent. That climbed to 3.5 percent as all workers focused on defeating fascism. In nationalized and war-related sectors the jumps were even more remarkable. For instance, in 1942, it took 3.2 worker hours to produce a pound of airframe. Three years later it took only 0.45 worker hours, one-sixth the time.

We can see these innovation and productivity effects already mirrored in the net-zero transformation underway. Rather than the costs of new technologies remaining stubbornly high (as many expected them to), once the diffusion S-curve steepens, aided by government expenditure, incentives, and regulation, costs fall rapidly, productivity rises, and the rate of innovation does not slip back but speeds forward. We see this in renewables, in batteries, and elsewhere. This dynamic of green industrial transformation can propel economies and our societies.

For these reasons, the GND and private sector investment shift could set the stage for a golden era in the US and globally, as in past post-war periods, of rapid, broad-based growth, falling inequality, and increased equality of opportunity. Real jobs with real wages may help change the conversation in locations and regions resistant to the new narrative. For example, more than 20 percent of job growth in Alabama, Alaska, Florida, Illinois, Kansas, Montana, North Dakota, and Wyoming in 2018 came from solar investment. This will begin to shift the nature of the local discussion to green facts underlined by economics. These positive economic dynamics are not just visible in the US but elsewhere as well, as businesses and investment shift and behaviours change. These are real investments in local communities across countries, not single apps downloaded from afar, enriching very few. Once this fact-based climate change story becomes embedded, this will only further support the transition.

Europe's green narrative shift underway

Europe's green growth is also dynamic and already productively rests upon a green narrative.

The EU has a 750 billion euro rebuilding plan and a 1 trillion euro GND, which Goldman Sachs calls the 'largest economic stimulus Europe has seen since the Marshall Plan' (CNBC, 2020). This is only the tip of the European GND, as Goldman Sachs estimates more than 7 trillion euros will be spent in the next decade on the plan. This is potentially transformational.

In a decade when Europe's growth has been fitful, anaemic, and disappointing, green jobs are outperforming others. Europe's environmental economy employment and value-added has outpaced the reset of the economy during 2000–2017, growing by 70 percent (Eurostat, 2018). The development of Europe's green economy is gathering pace. As the WEF notes, 'The programs offer the opportunity to reset the economy, create jobs, boost GDP and build resilience' (WEF, 2020). As Canfin (2020) notes:

> Europe could be the world's first carbon neutral continent by 2050. And if it balances the needs of both the climate and the economy, it could demonstrate how jobs and prosperity go hand-in-hand with environmental priorities. Europe can set an example for the rest of the world.

A reseeding and green regrowth is driving expansion in the EU small- and medium-sized enterprise sector, with 34 million EU jobs coming from the green sector, with services leading (8.6 million), followed by retail (7.8 million), industry (4.3 million), and manufacturing (3.2 million) (EU Open Data Portal, 2017). Europe's leading businesses understand this necessity in a way not yet fully internalized in the US or China.

Europe's policy and practices, public and private, are considerably in advance of the US and China. As the CEO of Action Group, a pro-green-growth business lobby primarily made up of European companies, has stated, 'We have to take more and faster action with more emphasis on sustainability and circularity. The European Green Deal presents an opportunity to do just this' (CEO Action Group, 2020). They are right.

European leaders use pointed, specific, and directive language in their calls and commitments to action. This is indicative that European CEOs understand this is an economy-wide transformation and that green is the growth of the future, brown investments must come to an end, disclosure requirements of the TCFD must become the norm, and carbon pricing and offsetting must be used. This shift in Europe is not just talk. It is increasingly and dynamically positive and environmentally progressive. Europe's policy and business narratives are greening and greener than elsewhere. This will continue to accelerate. The collective stories being told and retold in Europe are helping to green the policy conversation and the economy. They are not hampered by false narratives and anti-science bias to the same degree as in the US, or as blinkered as in Australia or Brazil. Serious public, civic, and policy circles do not entertain climate denialism. Others green stories are still under construction and face greater disputatious debate. Not so in Europe.

Japan signals that it, too, will leap

Japan, too, is making green changes to its industrial policy. Prime Minister Yoshihide Suga has promised to 'fundamentally shift' from coal to achieve net zero by 2020. In backing the net-zero goals, Suga will trigger actions across corporate Japan, which takes its lead from the government, and will move fast as the consensus gels. The government is moving to back solar power and CCS technology for emissions for various industrial applications. Here again, policy matters. With the government backing solar and CCS, the shift will accelerate across industry sectors. As the prime minister clearly stated, 'I declare we will aim to realize a decarbonized society. … Responding to climate change is no longer a constraint on economic growth. We need to change our thinking to the view that taking assertive measures against climate change will lead to changes in industrial structure and the economy that will bring about great growth' (climatechangenews, 2020). The prime minister understands the economic reality. Greening growth will be the new engine of Japan's future.

China, starting slower … but watch this space

China is beginning the green transformation, and it has a long way to go to bend the curve on GHG emissions to net zero by 2060. Sceptics should look at what China has achieved in the past before suggesting this is just greenwashing. China's ability to direct the organs of the state and business to common goals is staggering. There is increasing evidence that the policymaking power of China is being brought to bear on climate goals. As this becomes clear, the rate of change within China will accelerate. The country and its leaders understand they must invest massively today to lead the green economy tomorrow. They have already shown they can do it, for example on high-speed rail and solar PV. But they have only just begun. For instance, they have announced the planting of forests as a key goal and signalled they will reforest an area the size of Germany.[3] China always goes big on its policy transformations.

Johnson's Churchillian moment?

UK Prime Minister Boris Johnson is making the jump. Never one to avoid borrowing from elsewhere when it might be helpful, the prime minister has said he wants to see a 'fairer, greener and more resilient global economy' after Covid-19 and that 'we owe it to future generations to "build back better"' – lifting directly from US President Joe Biden's language (BBC, 2019). The UK prime minister has a personal responsibility. He is the host of COP26 and must help ensure it succeeds and that leaders together grasp the opportunity to force the break point with the past and speed our collective transition. Johnson needs to lead, not fumble and act like a buffoon. This is the prime minister's Churchillian moment (Mackintosh, 2020). He must seize it, but whether he has the personal and diplomatic skills to do so is not clear. Just as UK Prime Minister Gordon Brown rose to the occasion and led the G20 in its crisis response in 2008–2009, so this a test for the current prime minister, a biographer, incidentally, of Churchill. Can he do it?

Fifty shades of green

I have stressed that as this green narrative and economic transition accelerates, it ceases to be about green jobs and becomes about all jobs. The transition to 50 shades of green (supported by governments) will multiply and amplify the effects beyond the renewable energy industries and power sector across the global economy, in all sectors, as they align with new regulatory net-zero guardrail requirements. Green will become synonymous with what is productive, profitable, resilient, ethical, acceptable, forward thinking, and globally necessary and morally essential for our planet and our future. To conclude, GNDs and the growth they will seed and support are necessary to achieving the net-zero industrial and societal transformation. But what about critics who bemoan growth, who say less growth or no growth is the only solution?

Addressing degrowth

The concept of 'degrowth' was first suggested by Gorz (1972) and was built upon by Latouche (2009). It has been given its most clear iteration by Jackson (2011). Advocates of degrowth maintain that growth is uneconomic and unjust and that it is ecologically unsustainable and will never be enough. Degrowth supporters argue for a reorientation of our societies away from neoliberal notions of growth and constantly increasing consumption (and pollution and destruction) towards other measures and metrics of social and societal well-being. Jackson and Victor (2020) ask if we can address GHG emissions and climate change through a policy of degrowth.

Recent modelling by Jackson and Victor (2020) demonstrates the transition to net zero and stable sustainable prosperity is possible and achievable. Their model (which uses Canada) focuses on four areas necessary to achieve climate change goals: (1) electrification of the economy, (2) decarbonization of the electricity sector, (3) decarbonization of the non-electricity sector, and (4) non-carbon-related environmental improvements. It compares a base case (business-as-usual) scenario; a carbon reduction scenario, where government achieves 80 percent GHG reduction from 1990 levels by 2050; and a sustainable prosperity scenario with faster transition to net zero by 2040.

In the model, workers' income grows the fastest in the first scenario, from CAD\$57,000 to CAD\$100,000 in 2067, and in the second scenario per capita GDP reaches CAD\$92,000 by 2067. In the final scenario, income only rises from CAD\$57,000 to CAD\$65,000. Jackson and Victor stress that the first outcome should be avoided – since the environmental damage is severe, uncalculated for, and permanent. The second is a good result, and even the third radical approach still results in modest income increases. As they note, 'Conventional wisdom would suggest that such a transition is impossible without causing irreparable damage to prosperity and well-being in society. … this undesirable outcome is avoided' (Jackson and Victor, 2020: 7). The key takeaway from Jackson and Victor's (2020) work is as a rebuttal to those who doubt net-zero progress is economically possible, bearable, and achievable, before climate change disaster strikes. They conclude:

> The pursuit of economic growth at the expense of a deepening environmental crisis has a very high probability of catastrophe. On the other hand, there clearly are alternatives to this paradigm. For instance, substantial reductions in GHG emissions can be achieved without massive changes to the structure of society.
>
> *Jackson and Victor, 2020: 13*

Rebranding degrowth

The positive message is that we can achieve carbon reduction and net zero with only slightly slower growth in GDP in advanced economies. This is an argument to go further and even faster and still see modest growth, coupled with positive social, economic (higher wages, lower inequality, shorter working weeks), and necessary environmental outcomes. Unfortunately, this important message can often get lost because of the connotations that the 'degrowth' story contains and transmits.

Degrowth is not a successful narrative

As a political economy and electoral matter, degrowth remains a difficult sell. As a policy suggestion, the name alone ensures the concept is a negative and not a positive narrative, despite Jackson and Victor's modelling to the contrary. I have stressed how important the story and language we use are to outcomes and in constructing the new consensus for action.

Degrowth as a narrative construct is not going to work. The nuances of related and otherwise popular and sensible policies, such as higher wages, fewer working hours, and adjustments to working life, get lost in its negative shadow. Just as the 2020 calls in the US to 'defund the police' sent the wrong narrative message to citizens in the middle, so degrowth will split the conversation, not strengthen it. In addition, degrowth sounds too much like the already rich telling the poor to look forward to a worse future. Or at least it will be portrayed as such. Yet elements of Jackson and Victor's reimagining and models are positive and compelling. The nuances are lost because of a major narrative and language misfire.

Take what works, refashion it, and move forward

I suggest, therefore, we take what works from the degrowth proponents, that is, discussions around the nature of growth, the components of growth, how we gauge growth and its effects, and what should matter and what should not, and apply them in the context of net-zero responses, GND design, and mechanisms. Take Jackson and Victor's call for higher wages, affordable childcare, shorter work weeks, and other improvements to our well-being as workers within society. Many of these are eminently sensible and politically popular.

Fairness, trust, and opportunity

In many cases, progressive countrywide solutions of a form advanced by Jackson and Victor (2020) are already applied in the states most advanced in GHG reductions and social welfare provisions, demonstrating (in Sweden, the Netherlands, and Denmark) that you can have your greening social democratic cake and eat it too. It is not a coincidence that these countries have been most effective in their net-zero conversations, policy goals, and implementation stories. They are socially more inclusive, economically less stratified, culturally more cohesive, and less tribally split. One can cite the examples of the Netherlands; Scotland, my home; Finland; Norway; and Sweden. There are others, including US states, pushing forward as communities. These smaller, cohesive states and communities often practise a somewhat more maternalistic capitalism (Collier, 2018), which better protects people from the worst downsides of free markets through higher taxation. These states also operate with governments that are accessible and trusted and that deliver local and national services to all.

Successful states that demonstrate competence and develop trust have more room to pursue net-zero goals without undermining their goals of opportunity and relative equality of treatment, engagement, and burden sharing. Fairness matters as trust matters, for we are all together in the war on carbon. Voters need to be assured that the process of transition also addresses questions of equity and the need for a just transition. The green, better future cannot be reached and be sustainable if it is a creation of the rich and advanced that leaves the rest behind. Today, it is far from fair and balanced internationally.

Only when the burdens are seen to be fairer, locally and transnationally, will the collective consensus, the narrative of action, and the regreening of our collective commons be built upon and sustained. In constructing the contours of a green net-zero tomorrow, we need to ensure we do so not as *homo economicus* but rather as *homo economicus sympatico,* by recognizing the social and collective nature of our community and of the endeavour and our need for fairness and equity in the tasks ahead.

For the climate change burden, fairness is essential as we reseed and regrow

Fairness has been hardwired into human and nonhuman beings through evolution, and it is universal. Behavioural economics, child and adult psychology, and animal studies all show how important it is for survival. We all sense intrinsically what is fair and what is not. We are not *homo economicus* utilitarian actors. Rather, we recoil from unfairness and react negatively to it. This is known as the 'inequity aversion' (Heinrich, 2004). As humans, we watch closely how others are rewarded and react negatively if we feel we are rewarded less for the same effort.

Kahneman, Knetsch, and Thaler (1986), for example, demonstrated this in an experiment in which an individual is given US$10. They can decide how to divide this money with another person. If the other person agrees, the division is carried out. If not, both get nothing. What does the deciding person choose to do? How does the receiving person react? Standard economic theory suggests that the giver would try to keep as much as possible, and the receiver would take whatever is offered, as this is better than nothing. Not so. Most participants split the ten dollars evenly. Those who offer less are often rejected by the receiver and both get nothing. This is fairness in action. Better we get nothing than allow an exhibition of blatant unfairness and selfishness. Similar reactions are seen even in very young children.

Our evolved aversion to unfairness is also demonstrated in the work of McAuliffe, Blake, and Warneken (2017). Here, researchers asked children to play a simple game. Two children who do not know each other are paired up and are given an unfair distribution of candy. One of the two children – the decider – could accept or reject the allocation. If the decider accepts, both children get their candy. If the decider rejects it, both children get nothing. Imagine that the decider gets four and their partner gets one. What will they do? Researchers found that children frequently rejected the unfair advantage. They are willing to sacrifice their own rewards to prevent someone else from getting the short end of the stick. For the children, getting nothing is better than getting more than a peer, even a child whom they have just met. Our children thus turn out to often defend fairness for others. The same researchers staged another game, underscoring fairness is essential.

In this game, one child is again the decider, and this child decider keeps all the candy for themselves without sharing. Another child, an observer, has an option: They can intervene to stop the unfairness, but only at the cost of sacrificing some of their own candy in order to stop the witnessed unfairness. What does the observing child do? Do they intervene? It turns out that 'children regularly intervene, choosing to pay some candy to prevent the selfish decider from getting away with unfair behaviour' to another child (McAuliffe, Blake, and Warneken, 2017). So here again we see that even small children know what is not fair and will react to change the balance in favour of fairness.

The evolutionary preference for fairness and rejection of unfairness transcends humanity. Monkeys understand when they are being treated unfairly as well. Research by de Waal (2011) with capuchin monkeys demonstrates that when one monkey receives more than another for the same work – a grape instead of a measly slice of cucumber – this results in the ripped-off monkey being visibly upset and angry about the outcome, throwing away his cucumber reward in disgust. Even more strikingly, if the researcher rewards another monkey with a grape when they have done nothing to deserve the reward, the offended monkey often refuses to participate in the researcher's task any further. The monkey goes on strike (de Waal, 2011). Freeloaders are bad news and very upsetting for monkeys. They are terrible for humanity and for climate change as well.

Fairness and greater equity are necessary to secure and fund our planetary goals

Why make this digression into psychology, childhood conduct, and animal studies? Because tangible evidence of fairness and equity is and will be essential if we are to permanently change our collective narratives and secure net-zero goals.

As we have seen in the Covid-19 pandemic, heavy burdens can be borne in a health crisis, as they can in war. Much greater individual and collective burdens can be borne in the battle for planetary economic decarbonization and net zero by 2050, but this requires all of us to understand the story behind the challenges, to participate more equitably as countries, societies, communities, investors, and individuals. Freeloaders and selfish abusers of societal norms of fairness and equity in our climate change responses must be rejected and punished. For if the few freeload while too many others stagger under the weight, the climate change narrative and the response to it could be at risk due to the resistance of the least powerful who unfairly bear the largest burden.

In the end, the richer among us would do well to remember it is in our own interest to help sustain and ensure a more equitable social contract. Otherwise, when the balance is out of kilter, social unrest can and will result. People lose their commitment to their societies and politics if their wages stagnate and they see their lives getting harder and harder. They then seek out populist solutions. Eichengreen (2018) stresses that societal cohesion requires periodic adjustment of the social contract away from the rich and towards the majority and the poor. Eichengreen views this as a recognition that the elite thrive without revolution when they understand they must also provide and maintain the balance of the contract, not take more from the masses again and again, in a misguided meritocratic belief that they deserve ever greater rewards in a winner-takes-most society.

Time to rebalance

Our climate change break point and economic tipping point is a time to rebalance the social contract, address inequality and unfairness, and correct egregious abuses to restore community trust in our governmental institutions and common goals. We know what is fair. A monkey knows it. A child knows it. We can tell when we are being taken advantage of and being phished for phools (Akerlof and Shiller, 2015) or asked to work for less while the bosses prosper (Akerlof and Yellen, 1990).

Governments and societies should take a series of additional economic and regulatory steps to help ensure a greater degree of fairness, equity, and burden sharing, and foster climate change dialogue, adaption, mitigation, and response while enhancing societal cohesion.

A key step to addressing rising inequality is to close tax loopholes and erase freeloading opportunities and, in doing so, generate resources for collective action.

Address rising extreme inequality

To address matters of fairness as well as raise revenues, governments should look to tax policy changes that acknowledge and tackle rising extreme inequality in this, the new Gilded Age of conspicuous consumption and ultra-high net worth among the few. The world's 2,153 billionaires reportedly have more wealth than the 4.6 billion people who make up 60 percent of the planet's population (Oxfam, 2020). Such extremes are morally and ethically troubling and economically and societally destructive. Year after year, the billionaire and millionaire classes have taken control over a larger and larger proportion of the world's wealth, precisely at a time when the planetary, economic, and societal challenges are more fraught and increasingly urgent. Oxfam (2020) correctly notes:

> Governments created the inequality crisis – they must act now to end it. They must ensure corporations and wealthy individuals pay their fair share of tax and increase investment in public services and infrastructure.

Securing additional resources will help fund GNDs and the required transformation. To do this, governments and electorates must confront the inequality elephant in

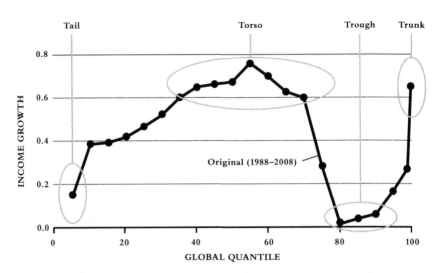

FIGURE 8.1 We must confront the income inequality elephant in the room

Source: Lakner and Milanovic, 2013.

the room. The Lakner and Milanovic (2013) inequality elephant graph (Figure 8.1) illustrates that the global elite, the top 1 percent, have seen massive income growth over recent decades that saw them capture a large share of global income growth. This can be seen in the elephant's raised trunk. The very rich are getting richer and richer, as income and compounding work in their favour.

The income of the global upper-middle class has stagnated, with zero growth over two decades for the 8th percentile. This is part of the explanation for the anger and electoral reaction of the working middle class, fuelling populist politics in advanced economies. This can be seen in the depth of the trough at the base of the elephant's trunk (Figure 8.1). The news has not been all bad – the total global middle class has risen rapidly as certain developing countries have grown, spurring the incomes of their populations. Countries like China have lifted large, impoverished populations into the middle class. This can be seen in the graph's peak at the elephant's torso.

The very poor have been left behind, mired in tragedy and struggle, with poorer countries stuck in a cycle of poverty and violence made worse by state collapse, itself affected by climate change dynamics and severe weather events. This can be seen in the elephant's slumped tail.

The story contained in Figure 8.1 is corrosive to societal stability, consensus, and the ability to respond to crises, including climate change. If there were greater equity and a sense of fairness in this and related resource transfers, the war on carbon could be more effective, acceptable, and not so disruptive of the existing order. If there are no steps to correct this elephantine problem, angry electorates may continue to stymie our common planetary goals, despite their manifest importance to us as a species.

As governments address climate change, countries should take steps to tackle the extreme wealth concentration among the very few and ensure that the wealthy pay a larger share of taxes. This is not a call to hose the rich, it is a call to modestly increase taxes paid by the very rich to levels still below what they paid as recently as 30 years ago.

Modest tax increases at the very top would result in a considerable contribution to government revenues. In the US, the Congressional Budget Office estimates that between 1979 and 2016, the income of the best-off 1 percent of American households nearly tripled before taxes. That same 1 percent of Americans hold slightly more wealth than the entire bottom 90 percent (Brookings Institution, 2019a).

Adjust taxation to address issues of equity and fairness

Governments funding GNDs should consider raising taxes on the wealth and incomes of the extremely rich. America provides an example of how this could rebalance equity and help finance the future. This could be achieved by:

- *Raising taxes on the extremely wealthy* by 2 percent on wealth over US$50 million and 3 percent for billionaires in the US. This minimal adjustment would raise up to US$2.75 trillion over ten years (Saez and Zuckman, 2014). This would be more than enough to pay for the Biden GND. Most US voters support this type of modest tax proposal.
- *Taxing unearned wealth* (via higher tax on corporate dividends). Ideally, governments should tax capital at the same rate as labour. As Piketty (2013) has made clear, absent policy action on taxes, capital will tend to accrue to fewer and fewer, richer and richer people, to the general detriment of society and its environmental stability.
- *Increasing estate taxes.* In the US, the top rate now only applies to those with estates of over US$11.2 million – less than one out of every thousand Americans. Only 1,900 Americans trigger the tax when they die. Only 80 small firms were affected by the estate tax. Brookings recommends governments should consider the merits of raising estate taxes to apply to estates valued at over US$3.5 million and graduating the rates above that threshold, a step that would recoup US$300 billion over ten years in the US.
- *Raising social welfare taxation.* In the US and in some other countries, there is a cap on social welfare taxes. If this tax were raised to affect 90 percent of earnings instead of the 83 percent today (in line with the level captured when the tax was created in 1930s), it would raise more than US$1 trillion over ten years (CBO, 2020).

These are just illustrative examples. Each country is different. Some countries have much higher tax rates (such as France, the Netherlands, and Sweden).

Others have much lower Gini coefficients (such as Australia and Japan). Others have fundamental problems of tax collection (such as Greece). The precise approach will vary country to country. The broad point is that governments, as they move to help underwrite and finance the green revolution, need to also rebalance the burden and address gross inequality. If inequality is left untreated, it might otherwise destabilize our societies and undermine the attempt to agree a common climate change narrative and secure our net-zero goals. I am not naïve about the probability of such big leaps, in terms of taxation of the very wealthy, as their hold on the levers of political power may be such that it is difficult to raise taxes back to reasonable, socially needed levels. Given that this is likely the case, other small, but still productive reforms are needed as we pursue a just transition.

Loopholes big enough to drive an SUV through

Governments should work to close egregious loopholes that are exploited by the wealthy few. After all, it is very hard to build a better tomorrow if the super-rich pay almost no taxes. For example, former President Trump paid no taxes for 10 of the last 15 years and only US$750 in taxes in the last two years, which is less than he paid in China or the Philippines (*New York Times*, 2020c). And yet he claims his net worth is in the billions. This is societally and morally unacceptable. If we are asking almost all workers in our societies to pay higher prices for carbon, to change their conduct and their eating habits, and to pay modestly higher taxes, we cannot permit the wealthy to continue aggressive tax avoidance to shirk societal and economic responsibilities.

Widespread tax avoidance and expensive state architecture and social programmes (let alone GNDs) cannot operate simultaneously in the medium and long term. Greece epitomizes this type of societal lack of trust and tax avoidance, with as much as a third of the economy operating outside the tax system. The Greek people and their governments have repeatedly lied about reality, and when caught they responded by prosecuting and hounding the country's chief statistician, Andreas Georgiou, for over a decade,[4] rather than admit their shame. Greek voters need to start a conversation about how they envisage their society paying for the future needs of the next generation if a third of their fellow citizens refuse to contribute anything at all. I do not know what the answer is in Athens, but surely a frank and civil conversation and debate is at least necessary, not to mention widespread and deep reforms, if the Greek state is to function, provide for its citizens, and prepared for and be able to transition to a net-zero economy.

Much more needs to be done to reform national tax codes in many countries to stamp out tax avoidance, and solve systemic bad behaviour, and address the actions of the super-rich. Further, in some countries, such as the US, corporations and the very rich have seen their taxes decline as public revenues fall and deficits rise. This avoidance and the associated loopholes are very costly, and greater effort should be taken to close the loopholes and capture lost revenues.

Apple's sour tax aftertaste

The tax avoidance case of Apple provides an illustration of egregiously aggressive international tax avoidance. Apple's scheme hinges on Ireland's sweetheart deal with the US-based company that allowed it to avoid Ireland's 12.5 percent corporate tax. Instead, Apple agreed to pay as little as 0.005 percent in taxes. Apple then syphoned all the profits the firm made in Europe through its Irish subsidiaries, paying this near-zero rate, instead of paying taxes where Apple products were purchased. In this way, Apple saved tens of billions of euros. A company that paints itself as a responsible actor turns out to be societal freeloader, a disdainful corporate libertarian failing in its corporate duty to the societies in which operates (Fair Observer, 2020). The European Commission has so far unsuccessfully sought repayment of these taxes.

Corporate giants that avoid their responsibilities

Scores of other major corporations, many with household names, also game the tax system and hold their profits in tax havens, like Apple does. Citizens for Tax Justice has ranked the top-30 Fortune 500 such companies (see Figure 8.2).

Other studies have come to similar conclusions. For instance, a Forbes analysis of Fortune 500 companies found that 60 were profitable yet avoided American federal income tax. The total US income of the 60, which includes giants such as Amazon, Chevron, Delta, General Motors, Haliburton, and IBM, was more than US$79 billion, but their effective tax rate was minus 5 percent. On average, they got tax refunds (*Forbes*, 2019). Governments should work to close these loopholes and coordinate internationally to require all firms pay a minimum rate of tax wherever they do business.

International coordination to avoid abuses

Governments should pursue international tax cooperation to shore up revenue flows and help pay for the essential green reseeding that is required. The OECD has been trying to close that massive gap for years. The world's leading nations should use that forum to stop avoidance. The OECD is seeking agreement on a minimum corporate tax rate on total revenue made in each jurisdiction, a reasonable move. The Biden administration should support this OECD endeavour; the Trump administration backed away from it after significant pushback from lobbyists for the firms affected. We need to pay for the green future; no one should get a free pass. This is fairness in action, and voters will understand this as long as it is explained clearly and repeatedly that these firms are damaging society by avoiding paying any taxes at all.

Here, as in so much else, we need a shift in the social narrative in countries where aggressive tax avoidance is seen as smart. It is not smart. It should be shameful. I recall when I worked for Mitsubishi Corporation, then the world's largest trading

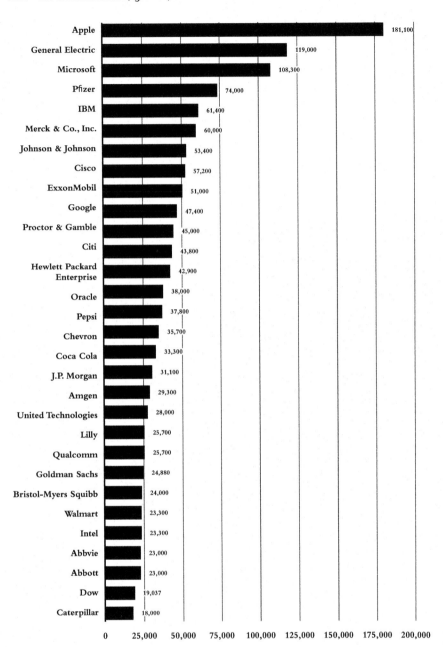

FIGURE 8.2 Corporate scofflaws should pay their fair share

Source: Citizens for Tax Justice, 2016.

company. The firm, as do others in Japan, pays taxes in Japan and the other countries where it operates. They do not aggressively avoid taxation. Japanese social norms require they pay taxes. Hence, the clear surprise from Chairman Ben Makihara, who contacted me one day to ask, 'I was having lunch with my good friend Mr Rockefeller last week, and he told me he pays no taxes. How can this be so?' A good question indeed. It should not be so. It should be legally, societally, commercially, ethically, and socially unacceptable to avoid paying taxes.

It is hard to discuss and address taxes in divided national political communities where the wealthy have an outsized influence and the poor have very little. Yet, solid majorities support tax fairness and oppose and disapprove of those who do not pay their fair share. So, there is possible democratic wiggle room for legislative man-oeuvre in favour of greater equity. There is no guarantee of success, but there is no sure-fire loss either. It is certainly worth repeated attempts and can be successful if the narrative and story is broadly understood, as it was in Denver.

It is also essential not to just deal with fairness in each country but also to tackle the thorny matter of redistribution of resources across borders. As a matter of sens-ible policy and human survival, we need strengthened mechanisms to speed the rate of green transformation among lower-income countries that need to be helped to leapfrog technologies. Advanced economies should facilitate these countries 'going straight to green'. The Paris Agreement made a start, but much more is needed.

Supporting the green transition, boosting common resource commitments

Governments understand resource transfers have to be part of the solution. It is untenable to require GHG cuts and goals while failing to support the transition among lower-income countries. Ultimately, in 2010, countries agreed to aim for a US$100 billion annual transfer by 2020, from rich to poorer countries, to support the leapfrogging process and rapid diffusion transition. Yet even this paltry figure has not been achieved, even though it is magnitudes lower than the actual need (Orenstein, 2015).

Double counting, miscounting, and other failures

States committed in principle to this (low) goal for resource transfers have engaged in questionable accounting to try and get to the annual US$100 billion figure, for instance, by counting overseas development assistance as being prin-cipally concerned with climate change, when this is not the case. Much of the total claimed and counted is not additional, just rebranded. This collective exercise in miscounting is disingenuous and does not help achieve GHG reduction goals. This is the not the first disappointment of the UN process and the gap between commitments and compliance and implementation. Nor will it be the last, as all too often the UN delivers as a forum for making commitments and speechifying but fails to ensure implementation and compliance. Disputes over how to get to

US$100 billion are fierce and have become more heated in the runup to COP26. Lower-income countries rightly complain about the persistent failure to support the transition and to match grandiose statements with new funds.

Let us assume for the moment that some unfortunate, creative diplomatic accountancy is used to claim that the US$100 billion figure has been reached by 2021. We should be sceptical if this happens. We should demand governments make much greater efforts through existing mechanisms and demand the funds are additive and effective. What do I mean? I believe we should be using the World Bank and regional MDBs to fund a new green Marshall Plan aimed at the net-zero transition.

A Global Future Fund or the 'Thunberg–Attenborough Plan'

Advanced economy leaders should take a small fraction of revenues and facilitate the lower-income countries' leap across technologies and innovations straight from their current level of development to green technologies. This should not be done via national overseas development assistance. The transition should be funded through a new, coordinated green Global Future Fund, perhaps called the Thunberg–Attenborough Plan, using World Bank and MDB financing mechanisms.

Just as we have dramatically increased World Bank and IMF resources to address the Covid-19 pandemic, so should leaders significantly increase climate change transition support by as much as an additional US$100 billion per year using existing multilateral institutional structures. This would amount to an approximately two-thirds increase in World Bank resources and a similar increase of US$50 billion spread across other regional MDBs.

Why take this route to provide more resources via existing mechanisms?

First, the cost is very low for the shareholders and creditors. The World Bank and MBDs do not require a dollar in for a dollar out, as they can use commitments and leverage to raise funds in the markets, as well as other mechanisms, which lowers the actual cost to shareholders while increasing the impact for borrowers and those receiving low-cost or no-cost grants. The current low for long interest rate environment, coupled to the triple A credit rating of the World Bank and MDBs, will continue to allow them to borrow at very low rates indeed and lend on at relatively low concessional rates to countries. While in the current environment some countries might prefer to borrow directly from the markets, the World Bank and MBDs can still provide pivotal support and steers on aspects of the green transition, while helping to ensure the projects that are backed achieve the social and environmental goals they claim to.

Second, the institutions, particularly the World Bank, have considerable green expertise. The bank represents the better nature of its shareholders, while being outside the direct individual control of single creditors. In addition, it understands development economics and has a track record of delivery. However, not all funds should come via Washington and the bank. To do so would concentrate too much power and overburden the bank, so a 50/50 split in new funds is proposed. This staff expertise advantage also applies for regional MDBs, where there is a balance among local knowledge, engagement, and oversight, and local buy-in. Thus, the Asian Development Bank has specific regional expertise, as does the Asian Infrastructure Investment Bank. In Europe and Eastern Europe, the EIB and the European Bank for Reconstruction and Development can apply commercial standards and speed the transition in that region, again at low cost to state shareholders.

Third, there are systems of governance already in place. Bilateral aid can be syphoned into favoured projects for political gain but which can be of little benefit to the recipients. The World Bank and MBD processes provide greater assurances that the funds are effective and are being overseen appropriately. Governance matters.

Fourth, if we believe the green transition should not be a top-down exercise alone but must also be a bottom-up and locally owned process, insulated from 'aid for trade' gaming of the system, then it is better to channel additional funds in a Global Future Fund via the World Bank and MDBs, rather than rely on double counting and ineffective, distorted national political mechanisms.

Finally, using existing institutions would allow leaders to stand up the additional funds faster, at very low cost. Additional resource transfers would have proper oversight and draw on expert communities. The bank and MDBs could set real GHG transition goals for the projects, as many such as the EIB are already doing in their other portfolios.

Is such collective greater fairness and generosity possible?

If we look at the current budgetary sleight of hand involved in attempts to come up with the Paris Agreement's number of US$100 billion, one might conclude a doubling of this flow of funds is impossible, but the Covid-19 crisis response suggests otherwise and shows us a great deal is in fact possible when all grasp the urgency and understand the drivers. Advanced countries are spending trillions of dollars annually on pandemic response and all understand this is not just possible but necessary. Are donor country populations willing to do more?

Donor country voters think they spend a lot on aid – but they don't

Polls repeatedly demonstrate two things about public perceptions and self-told stories about foreign aid that are important to making a judgement on whether greater support for resource transfers is or is not possible.

First, many voters overestimate how much is spent on foreign aid. For instance, Australians peg it at 14 percent of GDP. Survey participants are thus surprised to find it is in fact only 0.8 percent (Lowy Institute, 2020). Americans are even more misguided. Opinion polls consistently report that Americans believe foreign aid is in the range of 25 percent of the federal budget. In fact, at US$39.2 billion for the fiscal year 2019, foreign assistance is less than 1 percent of the federal budget (Brookings Institution, 2019b).

Second, often voters are more generous than we believe. Thus, when asked, Australians surveyed responded that a figure of 10 percent would be acceptable for foreign aid. When asked how much US aid should be as a percentage of spending, Americans also say about 10 percent. This generous mindset is also seen in Germany and France, where 35 percent support increasing overseas development assistance, even as the French often think the money can be wasted.

We are not all irredeemably selfish; in fact, as mentioned, evolutionary drivers make us cooperators and monitors of what is fair and what is not. We worry about outcomes and bad actors. This general evolutionary fairness and generosity of peoples, together with the apparent confusion over levels and actual amounts (on the upside), provides an opening to start a new conversation.

A new conversation based on facts and common objectives

Here is an apparent opening for a new narrative on aid and the green transition. It can be one constructed via dialogue and based on the facts of the very modest current and proposed level of support. It can be a story drawing on voters' better natures while stressing the low cost of such a plan, which is far less than many believe is reasonable, affordable, and appropriate. This narrative should stress not only the crucial climate benefits of the very modest support, but also its economic multiplier effects. Most of all, we need a narrative on support of other communities that counters rumour and falsehoods with the benign and positive reality and a cost that is even lower when we use existing institutions that can borrow in the markets and thus allow us to achieve our global commons goals, even if we double them from the US$100 billion currently agreed.

National discussions and conversations should be firmly grounded in the facts, and in this manner construct a consensus on modestly more generous and effective levels of support, directed through mechanisms we know work and already exist.

There is no guarantee of success in shifting our resources story, but it should be attempted. Remaking the story of aid and our common goals needs to be accompanied by making commitments real. It is no good to engage in double counting, reallocation, and wishful thinking. That does not help; rather, it breeds cynicism on all sides – disappointment and cynicism on the donor side, with donor electorates seeing the impact on GHG emissions and concluding the aid is being misused; and upset on the receiving side, who know whether funds are additional or just the same numbers being called something else.

A coming resource stress test for leaders

The resource question is a further diplomatic stress test for COP leaders. Setting aside the need for even greater resources, better used and funnelled, as I have suggested, will the current US$100 billion figure be reached, and will it be real or not? If yes, we are possibly brought slightly closer to a consensus deal in other areas on offsets, pricing, targets, and so on. If not, we have a gap between the polluters' demands for action, which are legitimate but stained by their own past, and developing countries' reasonable demands for modest support to leapfrog technologies and join the transformation process.

G20 leaders need to invest in our common futures. They need to look over the political horizon to the climate change reality that is unfolding and move collectively to address it before it is too late. The price is small, and the overall possible benefit to the planet colossal.

Let the transition begin in earnest

The climate change transformation to net zero is underway. Indications of the shift are increasingly visible in public commitments, plans, goals, and strategies. More and more private sector actors are on board and pulling in the same direction, with positive feedback loops between public and private actions and public communities, and within private businesses. Many leading governments have made aggressive new net-zero commitments and have launched GNDs.

COP26 is the next test – the next step change – in the process. It is a chance to support, confirm, and reward emerging market dynamics; to speed the rates of diffusion, innovation, mitigation, and adaption; to invest in a green future that will come either in a managed way or in a disorganized, destructive way; and to begin to restructure and create a just transition.

We are in a race of tipping points from one equilibrium to another. COP26 offers a chance to get much closer to a planetary policy tipping point – to common national societal tipping points to the green transition. If we succeed in getting to our own narrative and policy tipping points, we pull net-zero closer and may avoid triggering planetary climate change tipping points of no return.

We know that such changes happen slowly, and then suddenly. Let us hope we are about see a sudden narrative and policy tipping point in 2021.

Notes

1 See www.wemeanbusinesscoalition.org/net-zero-2050.
2 See https://transformtonetzero.org.
3 China is 27 times larger than Germany.
4 Andreas Georgiou was the chief statistician of the Greek government at the time of the Eurozone and Greek crises. When he quickly discovered massive underreporting of government debt, he reported the correct figures to the EU and IMF. For reporting the facts, Mr Georgiou was repeatedly and maliciously prosecuted all the way to Greece's supreme court by a corrupt establishment manifestly unwilling to own up to their total failure. Mr Georgiou eventually secured a measure of success, with the EU, IMF, and other governments demanding an end to this attack on facts and a hardworking civil servant. But Mr Georgiou was forced to live outside his own country for more than a decade, a terrible story of government abuse of the judicial and legal processes.

References

Akerlof, G. and Shiller, R. (2015) *Phishing for Phools: The Economics of Manipulation and Deception.* Princeton: Princeton University Press.

Akerlof, G. and Yellen, J. (1990) 'The fair wage-effort hypothesis and unemployment'. *Quarterly Journal of Economics*, 105 (2) (May): 255–283 [Online]. Available at: www. washingtonpost.com/blogs/wonkblog/files/2013/10/fair_wage_effort_hypothesis.pdf (accessed: 30 November 2020).

Andres, L., Couberes, D., Diouf, M.A., and Serebrisky, T. (2007) 'Diffusion of the Internet: A Cross-Country Analysis,' Policy Research Working Paper 4420, World Bank, Washington, DC [Online]. Available at: https://openknowledge.worldbank.org/bitstream/handle/10986/7636/wps4420.pdf?sequence=1 (accessed: 1 February 2021).

Atlantic. (2019a) 'The green new deal's big idea', 19 February [Online]. Available at: www.theatlantic.com/science/archive/2019/02/green-new-deal-economic-principles/582943/ (accessed: 1 February 2021).

BBC. (2019) 'Demands grow for "green industrial revolution"', 3 June [Online]. Available at: www.bbc.com/news/science-environment-52906551 (accessed: 30 November 2020).

Brookings Institution. (2019a) 'Who are the rich and how might we tax them more?', October [Online]. Available at: www.brookings.edu/policy2020/votervital/who-are-the-rich-and-how-might-we-tax-them-more (accessed: 12 November 2020).

———. (2019b) 'What every American should know about US foreign aid', 2 October [Online]. Available at: www.brookings.edu/opinions/what-every-american-should-know-about-u-s-foreign-aid (accessed: 16 November 2020).

Canfin, P. (2020) 'How the EU Green Deal offers hope for European recovery', 19 October. Microsoft blogs [Online]. Available at: https://blogs.microsoft.com/eupolicy/2020/10/19/how-eu-green-deal-offers-hope-for-european-recovery/ (accessed: November 2020).

CBO (Congressional Budget Office). (2020) 'Increase the maximum taxable earnings for the social security payroll tax', 9 December [Online]. Available at: www.cbo.gov/budget-options/56862 (accessed: 1 February 2021).

CEO Action Group. (2020) 'Business outlines plan of action to accelerate the European Green New Deal and post-COVID-19 recovery, engagement material'. World Economic Forum, September [Online]. Available at: www3.weforum.org/docs/WEF_Joint_Statement_of_the_CEO_Action_Group_for_the_European_Green_Deal_2020.pdf (accessed: 17 November 2020).

Citizens for Tax Justice. (2016) 'New report exposes world of offshore tax avoidance.' 5 October [Online]. Available at: https://ctj.org/new-report-exposes-world-of-offshore-corporate-tax-avoidance (accessed: 22 March 2021).

climatechangenews. (2020) 'Japan net zero emissions pledge puts coal in the spotlight', 26 October [Online]. Available at: www.climatechangenews.com/2020/10/26/japan-net-zero-emissions-pledge-puts-coal-spotlight (accessed: 30 November 2020).

CNBC. (2020) 'Goldman Sachs picks 20 stocks to ride Europe's push toward a greener future',10 July [Online]. Available at: www.cnbc.com/2020/07/10/goldman-sachs-picks-20-stocks-to-ride-europes-push-toward-a-greener-future.html (accessed: 1 February 2021).

Cohen, S. and Delong, B. (2016) *Concrete Economics.* Cambridge, Massachusetts: Harvard Business Review Press.

Collier, J. (2018) *The Future of Capitalism: Facing the New Anxieties.* New York: Harper.

de Waal, F. (2011) 'Moral behavior in animals'. TED Talk, November [Online]. Available at: www.ted.com/talks/frans_de_waal_moral_behavior_in_animals?language=en (accessed: 11 November 2020).

Eichengreen. B. (2018) *The Populist Temptation: Economic Grievance and Political Reaction in the Modern Era.* Oxford: Oxford University Press.

EPI (Economic Policy Institute). (2020) 'Investment in infrastructure and clean energy would create at least 6.9 million good jobs', 20 October [Online]. Available at: www. epi.org/press/investment-in-infrastructure-and-clean-energy-would-create-at-least-6-9-million-good-jobs-manufacturing-construction-and-transportation-sectors-would-gain-the-most (accessed: 1 February 2021).

EU Open Data Portal. (2017) 'Flash Eurbarometer No. 456: SMEs, resource efficiency and green markets' [Online]. Available at: https://data.europa.eu/euodp/en/data/dataset/S2151_456_ENG (accessed: 25 January 2021).

Eurostat. (2018) 'Development of key indicators for the environmental economy and the overall economy, EU-27, 2000–2017' [Online]. Available at: https://ec.europa.eu/eurostat/statistics-explained/index.php?title=Environmental_economy_%E2%80%93_statistics_on_employment_and_growth (accessed: 17 November 2020).

Fair Observer. (2020) 'Only losers pay taxes: Apple and the ingenuity of tax avoidance', 24 July [Online]. Available at: www.fairobserver.com/economics/hans-georg-betz-apple-tax-avoidance-verdict-ireland-eu-tax-havens-news-18811 (accessed: 11 November 2020).

Fink. L. (2021). 'Larry Fink's 2021 letter to CEOs.' [Online]. Available at: www.blackrock.com/us/individual/2021-larry-fink-ceo-letter (accessed: 22 March 2021).

Forbes. (2019) 'Renewable energy job boom creates economic opportunity as coal industry slumps', 22 April [Online]. Available at: www.forbes.com/sites/energyinnovation/2019/04/22/renewable-energy-job-boom-creating-economic-opportunity-as-coal-industry-slumps (accessed: 16 November 2020).

Georgson, L. and Maslin, M. (2019) 'Estimating the scale of the US green economy within the global context'. *Palgrave Communications*, 5 (121). https://doi.org/10.1057/s41599-019-0329-3 (accessed: 25 January 2021).

Gordon, R. (2016) *The Rise and Fall of American Growth: The US Standard of Living since the Civil War.* The Princeton Economic History of the Western World. Princeton: Princeton University Press.

Gorz, A. (1972) *Nouvel Observateur*, 397, 19 June. Proceedings from a public debate organized by the Club du Nouvel Observateur, Paris, France.

Heinrich, J. 2004. 'Inequity aversion in capuchins?'. *Nature*, 428 (139) [Online]. Available at: https://doi.org/10.1038/428139a (accessed: 1 February 2021).

Jackson, T. (2011) 'Prosperity without growth: Economics for a finite planet'. *Energy & Environment*, 22 (7): 1013–1016. Available at: https://doi.org/10.1260/0958-305X.22.7.1013 (accessed: 11 February 2020).

Jackson, T. and Victor, P. (2020) 'The transition to a sustainable prosperity: A stock-flow-consistent ecological macroeconomic model for Canada'. *Ecological Economics*, 117: 1–14 [Online]. Available at: https://reader.elsevier.com/reader/sd/pii/S0921800920301427 (accessed: 10 November 2020).

Jotzo, F. (2016) 'Decarbonizing the world economy'. *Solutions*, 7 (3): 74–83 [Online]. Available at: https://issuu.com/thesolutionsjournaldigital/docs/fea_jotzo (accessed: 16 February 2021).

Kahneman, D., Knetsch, J.L., and Thaler, R.H. (1986) 'Fairness and the assumption of economics'. *Journal of Business*, 59 (4) Part 2: S285–S300.

Lakner, C. and Milanovic, B. (2013) 'Global income distribution: From the fall of the Berlin Wall to the great recession', Policy Research Working Paper Series 6719, World Bank [Online]. Available at: https://ideas.repec.org/p/wbk/wbrwps/6719.html (accessed: 16 November 2020).

Latouche, S. (2009) *Farewell to Growth*. Cambridge, UK: Polity Press.

Levitz, E. (2020) 'The economy of World War II proved a Green New Deal is possible', 24 September [Online]. Available at: https://nymag.com/intelligencer/2020/09/biden-climate-green-new-deal-world-war-two.html (accessed: 1 February 2021).

Lowy Institute. (2020) 'Lowy Institute Poll 2020' [Online]. Available at: https://poll. lowyinstitute.org/themes/foreign-aid (accessed: 16 November 2020).

Ludwig, G. (2021) Ludwig Institute for Shared Economic Prosperity. [Online]. Available at: www.lisep.org (accessed: 18 February 2021).

Mackintosh. S. (2020) 'The Great British kakistocracy'. *EuropeNow*, 20 November [Online]. Available at: www.europenowjournal.org/2020/11/20/the-great-british-kakistocracy (accessed: 1 February 2021).

Mason, J. W. and Bossie, A. (2020) 'Public spending as an engine of growth and equality: Lessons from World War II'. Roosevelt Institute, 23 September [Online]. Available at: https://rooseveltinstitute.org/publications/public-spending-as-an-engine-of-growth-and-equality-lessons-from-world-war-ii (accessed: 1 February 2021).

Mazzucato, M. (2019) 'The economic argument behind the Green New Deal'. *MIT Technology Review*, 24 April [Online]. Available at: www.technologyreview.com/2019/04/24/135779/the-economic-argument-behind-the-green-new-deal (accessed: 10 November 2020).

McAuliffe, K., Blake, P.R., and Warneken, F. (2017). 'Do kids have a fundamental sense of fairness?', 23 August [Online]. Available at: https://blogs.scientificamerican.com/observations/do-kids-have-a-fundamental-sense-of-fairness (accessed: 11 November 2020).

New York Times. (2020a) 'GM drops its support for Trump climate rollbacks and aligns with Biden,' 23 November [Online]. Available at: www.nytimes.com/2020/11/23/climate/general-motors-trump.html?action=click&module=Spotlight&pgtype=Homepage (accessed: 23 November 2020).

———. (2020b) 'GM accelerates its ambitions for electric vehicles', 19 November [Online]. Available at: www.nytimes.com/2020/11/19/business/gm-electric-vehicles.html (accessed: 22 November 2020).

———. (2020c) 'Trump paid $750 in federal income axes in 2017. Here's the math', 27 September [Online]. Available at: www.nytimes.com/2020/09/29/us/trump-750-taxes. html (accessed: 17 November 2020).

OECD (Organisation for Economic Co-operation and Development). (2017) 'Employment implication of green growth: Linking jobs, growth, and green policies', June [Online]. Available at: www.oecd.org/environment/Employment-Implications-of-Green-Growth-OECD-Report-G7-Environment-Ministers.pdf (accessed: 17 November 2020).

Orenstein, K. (2015) 'COP Blog: Paris's $100bn question'. EnvironmentalFinance.com, 1 December [Online]. Available at: www.environmental-finance.com/content/analysis/cop-blog-pariss-100bn-question.html (accessed: 16 November 2020).

Oxfam. (2020) 'World's billionaires have more wealth than 4.6 billion people', 20 January [Online]. Available at: www.oxfam.org/en/press-releases/worlds-billionaires-have-more-wealth-46-billion-people (accessed: 12 November 2020).

phys.org. (2019) 'US green economy worth $1.3 trillion per year, but new policies needed to maintain growth', 15 October [Online]. Available at: https://phys.org/news/2019-10-green-economy-worth-trillion-year.html (accessed: 16 November 2020).

Piketty, T. (2013) *Capital in the Twenty-First Century*. Cambridge: Harvard University Press.

Rodrik, D. (2017) *Straight Talk on Trade: Ideas for a Sane World Economy*. Princeton: Princeton University Press.

Saez, E. and Zuckman, G. (2014) 'Wealth inequality in the United States since 1913: Evidence from capitalized income tax data', NBER Paper 20625, National Bureau of Economic Research, Cambridge [Online]. Available at: http://goodtimesweb.org/industrial-policy/2014/SaezZucman2014.pdf (accessed: 4 January 2020).

Scheidel, W. (2018) *The Great Leveler: Violence and the History of Inequality from the Stone Age to the Twenty-First Century*. Princeton: Princeton University Press.

Schwartz. A. (2020) 'Who needs a just transition?'. Center for Strategic and International Studies, 21 May [Online]. Available at: www.csis.org/analysis/who-needs-just-transition (accessed: 20 January 2021).

Summers, L. (2014) 'Reflections on the new "Secular Stagnation hypothesis"'. VoxEU, 30 October [Online]. Available at: https://voxeu.org/article/larry-summers-secular-stagnation (accessed: 18 February 2021).

UNFCC (United Nations Framework Convention on Climate Change). (2020) 'Commitments to net zero double in less than a year', 21 September [Online]. Available at: https://unfccc.int/news/commitments-to-net-zero-double-in-less-than-a-year (accessed: 30 November 2020).

WEF (World Economic Forum. (2020) 'Business leaders embrace Europe's new green reality for investment and growth', 16 September [Online]. Available at: www.weforum.org/agenda/2020/09/business-leaders-embrace-europe-s-new-green-reality-for-investment-and-growth (accessed: 17 November 2020).

9

A RACE OF TIPPING POINTS

We are in a race of tipping points on climate change. We need to reach a collective narrative policy and practice tipping point before we hit a series of terrifying, interlinked climate change tipping points that cause irreversible climate breakdown and that lead to a hothouse world from which we can never return. This race of tipping points is the epochal planetary challenge of our time.

We know from the data that, as Mann states:

> The more observations we get, the more sophisticated our models become, the more we're learning that things can happen faster, and with a greater magnitude, than we predicted just years ago.
>
> *Mann, 2019*

Success in achieving the narrative and policy leap requires massive changes across our economies, societies, and communities. If we fail, through lack of leadership, failure to stretch, or unwillingness to change, the prosperity and survival of all societies will be at risk. Not only do we face disaster if we fail, but our inaction will also cause the mass extinction of a large proportion of nonhuman species on the planet.

In 2021, a great deal is at stake, yet we appear far from the necessary narrative and policy tipping point. Most states continue to fail to deliver on their net-zero goals. Most industrial sectors have still to decisively shift production processes to be carbon neutral or circular. Meanwhile the GHG stock is building, and our global carbon budget is almost exhausted. Should we therefore lose hope, crushed by despair that the problem is too large or too complex to be solvable?

No. Now is not the time for hopelessness. There will be time enough for that if we fail. Just as Mann (2019) observes, the rhythm of the climate crisis is slow and

DOI: 10.4324/9781003037088-10

then very fast; narrative, policy, business, and societal shifts operate in such cycles. Our stories and narratives appear solidly resistant, until they suddenly and decisively shift once the consensus shifts. When this happens, a great deal can change, and change fast.

We have seen this process of sudden shifts play out in the pandemic response and reaction. We see again that in a crisis, leadership matters – that you must act now and not delay. We see that once a majority recognizes and understands a crisis, a great deal is possible. New coalitions of the willing form. Old beliefs are cast aside. The previously impossible becomes possible. Huge reserves of state power and authority can be brought to bear. We saw that when crises are recognized and understood, people, communities, and families pull together and are willing to change their conduct, behaviours, and expectations for the short, medium, and even long term. Economies and businesses also change swiftly and reorient their approaches and models. The pandemic we are living through illuminates that crises are crucibles of narrative, policy, politics, economics, and societal change.

Facing the climate change emergency, we are engaged in constructing a new collective understanding, a climate crisis economics story, a political economy response that is resilient, sustainable, and ethically and morally defensible.

We can win this race of tipping points. There is still time – if we learn from the pandemic and other crises. In 2021, there are signs that the world may be approaching a series of interlinked and reinforcing narrative, policy, and economic tipping points that could set the stage for Green Globalization 2.0.

Leadership always matters

Leadership always matters in policymaking and implementation. Addressing the climate change crisis requires farsighted, altruistic, ethical, and dynamic leadership by our political class. Global leaders coming together at COP26 in November 2021 must, through their actions, declarations, and commitments, help us reach our narrative tipping point on the climate crisis urgency, turbocharge their responses, and plan the route ahead. In doing so, leaders can decisively affect economic and market expectations and sentiment and confirm the evolving green story that is developing. Leaders can signal by their collective will to act that the glidepath to net zero is a policy goal across all policy areas. There is growing evidence this governmental leap may be taking shape in 2021.

President Biden recognizes the urgency and need for action and has stated:

> We've already waited too long to deal with this climate crisis. We can't wait any longer. …We see it with our own eyes. We feel it. We know it in our bones. And it's time to act.
>
> *Biden, 2021*

John Kerry, President Biden's climate czar, also made clear the urgency of now, calling on world leaders to:

> Treat the crisis as the emergency that it is. … We've reached a point where it's an absolute fact that it is cheaper to invest in preventing damage or minimizing it at least than cleaning up. … we have to mobilize in unprecedented ways to meet this challenge that is fast accelerating, and we have limited time to get it under control.
>
> *Kerry, 2021*

To respond effectively to a crisis requires recognizing it is such. Both President Biden and Mr Kerry clearly have made this leap, which then leads to policy shifts and action. Had Trump won in 2020, this book would have had a very different tone and a depressing conclusion. With the Biden administration placing climate change at the top of its national security and diplomatic agenda, the global narrative and policy pathways are open and widening. The US administration has already announced a raft of policy changes and unveiled a US$2 trillion green industrial policy. This is not greenwashing. This is a crucial narrative and policy tipping point.

Other leaders preparing for COP26 have increased the level of their ambition.

Act now

Major states, including the US, China, Japan, and the EU, have committed to net-zero goals. Many others must do likewise. In making such commitments, the collective story on climate change can shift dramatically and permanently.

China has made a leap and launched its net-zero 2060 drive. The details of how it will be implemented must be scrutinized closely. President Xi Jinping needs to lay out in much more detail the steep glidepath and waystations, before COP26. China's next five-year plan will be the real indicator of national policy, political urgency, and the extent to which China will swiftly align with net-zero goals. The initial indications are positive. For instance, China has announced a goal to get 20 percent of primary energy from non-fossil-fuel sources by 2025. This translates into making 42 percent of China's grid renewable- or nuclear-powered, up from about 32 percent in only five years. This is a huge increase, but analysts believe China can achieve it and go still further after 2025. This is only one commitment. There are others, for example, on carbon taxation, ETSs, and reforestation. It is increasingly clear that China understands the green industrial revolution has begun and wants to lead, not lag. This is good news. Without China's net-zero narrative, the GHG goals of Paris were out of reach. Now, the future looks a bit brighter. Having China and the US both on board and making narrative and policy leaps towards net zero is essential. This changes the climate calculus from one of despair and tragedy to one of possibilities and opportunities. With the two economic superpowers behind decarbonization, the outcome, and our planetary story, can shift and, potentially, rapidly change.

Europe, the UK, and Scotland are already lighting the way forward, with the most aggressive, sustained, and effective reductions in GHG emissions in the advanced world. Policymakers in Europe have understood the climate change narrative for years. Europe's leaders are not playing catch up as the US and China are; European states are already applying many of the policy levers and mechanisms needed to shift incentives. Others should learn from their successes and failures.

The national and international climate change glidepaths ahead will be steep, challenging, and turbulent, but they will be steeper still if we delay. The plans and implementation will face political headwinds. But increasingly, denialist demands are being drowned out by the majority (voters and businesses) who support action and want it now. Many people understand that the time to invest in a green tomorrow is today.

Stop discounting the future

As John Kerry said, governments must stop discounting the future. Getting to net zero requires that we stop discounting the value and survival of future generations. From an economic but also a moral and ethical perspective, this practice is unacceptable and unsustainable. From a planetary perspective it is unforgiveable. For too long, discounting has acted as a brake on necessary action. It needs to stop. Governments should slash the discount rate for climate change investments to as close to zero as possible. A massive, sustained, transformational, green investment boom should be financed by governments, the private sector, and investors.

The necessary investments can be brought forward by a near-zero discount rate. Moreover, with interest rates extremely low or even negative across much of the advanced economies, the cost of green investment is negligible, while the positive feedback loops in the economy will be significant, broad-based, and lasting. Green government and private investment in rebuilding, retrofitting, and redesign should be pursued on a massive scale. In 2021, governments in the US, China, Europe, and elsewhere appear to recognize the economic and political economy sense behind a prolonged investment surge underpinned by green industrial policies and national and business strategies aligned with marking to planet and our climate goals.

The multiplier effects of decades of investment in the construction of and transition to a Green Globalization 2.0 are clear. Investment policies prudently pursued and supported by governments and the private sector will pay real and persistent economic, societal, and planetary dividends in the decades ahead. Already markets recognize this shift and are rewarding first movers and leaders.

As we stop discounting and invest in our green tomorrows, we must start charging the real price for carbon and further shift market, commercial, and individual incentives once and for all.

On pricing carbon, convergence, and shifting incentives

Scores of states have begun the process of pricing carbon through taxes or cap-and-trade schemes. The example set by Sweden, the EU ETS, and Canada, for instance, illuminate how governments can set aggressive goals and design the needed glidepath. In 2021, a rapid transition to pricing carbon at an agreed minimum, increasing progressively and converging globally, is essential. That is economics 101. We know pricing changes market conduct, pulling forward actions, shifting decisions and business and personal choices. The US SO_2 market showed us how to do it and achieve atmospheric goals. We also know from the example of the few states that have commenced the war on carbon that it is manageable and economically beneficial. Pricing carbon requires political guts and clarity, and consideration of issues of equity and fairness, but it can be done without huge disruption and can result in major GHG emissions reductions, as Sweden, Canada, and others have demonstrated.

This is a crucial litmus test of global leaders. Can they converge on and agree on carbon pricing goals and an upward trajectory? Can they announce a 'C-day' and prepare for it? Leaders should reach for much higher carbon pricing than seen at present in most markets, converging around US$130-plus a ton in 2030 and rising to US$300 by 2050. We need to confront this economic and market failure to price carbon head-on. The precise price is a matter for debate and some compromise, but two general requirements are indisputable: (1) a carbon price minimum needs to be agreed, and (2) the carbon price should then progressively increase in the decades running up to 2050. These two steps are prerequisites to create predictability, enhance credibility, shift market expectations, and pull forward investment decisions and the rate of industrial change and green innovation.

On the crucial matter of carbon pricing, I am pessimistic. The COP process is not well suited to making such leaps, however much the planet requires it. The probability of a few states holding out against progressive carbon pricing is high. One can imagine Russia, Saudi Arabia, and other carbon-addicted states refusing point-blank to agree a meaningful pricing plan and accord. What should be done if no consensus of a 'C Day' can be reached?

Time for a coalition of the willing

Leaders who are pricing carbon and raising prices must construct a coalition of the willing. States that take on the responsibility of planning for and executing a sustainable and resilient net-zero pathway must not be undercut by freeloaders, polluters, and denialists. Net-zero plans are not cost free. Neither is carbon pricing. States and the firms that commit and execute on agreed climate change plans must be protected from those that do not. In 2021, leaders and states ready to move ahead must do so regardless of those who refuse to act. Leaders who are pricing carbon should therefore tax the goods of those who refuse to protect the planet.

Time to tax the freeloaders to force the pace of change

A coalition of the willing must form if a strong global consensus on net zero and carbon pricing does not coalesce. We need to tax the freeloaders to force the pace of GHG reductions. This coalition should use a carbon border tax to stop freeloaders' firms from undercutting and unfairly competing against those that carry the burden of pricing carbon. We cannot allow carbon dumping. Neoliberal proponents of a version of free market economics that does not price carbon or value the plenary ecosystem that we live within will strongly object. We should close our ears to such histrionics. Leaders need to ensure that the countries, firms, and markets that are engaged in the mammoth and critical task of securing a liveable planet do so without being undermined by denialists. Polluters must pay a high price and so be pressured to reconsider and join the net-zero consensus and coalition.

It is time to support market movements to signal a climate narrative break point

Bold action in Glasgow at COP26 and beyond must support changes in market sentiments, stories, and currents that are already visible in the environmental, social, and governance space, among young investors, in certain industrial sectors, and among leading dynamic firms that are seizing green opportunities now. Leaders who take the leap during COP26 would not be in advance of market currents but would be riding on them, building them up across economies, sectors, and industries. Decisive action in Glasgow can reconfirm and accelerate the rate of shift.

Markets are on the move. Leaders should harness them and encourage them, pull forward investment decisions, shift strategies, and alter individual conduct and choices. By doing so, governments would amplify the effect of their policy shift, leverage the impact of their investments, and reconfirm to investors that green investment, in all its 50 shades, is a good choice. In doing so, governments can widen the gap between brown and green, push the smart firms in the former industries to alter strategies and business plans, reward the leaders, and spur the laggards to shape up and shift plans and investments.

On building institutions to oversee our greening economy

As governments announce aggressive goals, confirm net-zero glidepaths, and set pricing and market expectations, leaders should also plan for new mechanisms to ensure commitments are delivered, targets are reached, and enforcement is fair and measured. GHG reduction and carbon pricing must be consistent and stringent, with effective implementation and enforcement.

Leaders need to reach a consensus on how they will globally coordinate pricing regimes, ensure fairness, and enforce compliance. We know regimes without compliance mechanisms do not work, internationally or nationally. Conversely, pollution pricing schemes with enforcement, from the US SO_2 market to the reformed EU ETS to the California ETS, deliver GHG reductions and alter market incentives.

Leaders need a global institution to police Green Globalization 2.0. They should create a World Carbon Organization (WCO) that can work in parallel with and learn the lessons from the WTO architecture. Ultimately, a smoothly operating Green Globalization 2.0 requires constancy, clarity, transparency, and enforcement of carbon pricing. A WCO can help ensure consistent convergence on carbon pricing and play a role in taxing freeloaders and polluters. Creating a WCO to apply quasi-judicial oversight and rulings on carbon pricing regimes and their comparability and application to net-zero goals is the right way to proceed. Governments need to avoid widely disparate mechanisms, regulations, carbon pricing measures, and spotty application. A spaghetti bowl of carbon pricing and regulations would not deliver GHG reductions smoothly or fairly.

Leaders in Glasgow will not make this architectural leap. But the need for a WCO will become increasingly apparent. As states apply carbon pricing and strengthen Green Globalization 2.0, they will need an institution to ensure compliance and comparability. The WCO construct is a way towards consistent international carbon pricing regulation.

Just as leaders need an international forum to ensure trade and the transition operate smoothly and achieve common climate change and decarbonization goals, so countries also require strong national oversight mechanisms. We have seen how some states have constructed independent carbon councils to hold their governments to account, to measure their net-zero progress against stated commitments. The UK, France, Ireland, New Zealand, and Sweden have all created such bodies. All other states need similar constructs, which I call National Carbon Banks (NCBs).

Politicians must agree to net-zero goals and legislate to achieve the goals. Then they should delegate the responsibility for reporting on progress, critiquing policy, and recommending pricing and supply changes to a technocratic agency modelled on central banks. By creating NCBs, leaders can leverage technocratic credibility, enhance predictability, and strengthen communication on goals and policies, and where necessary recommend and (ideally) require pricing changes and market oversight adjustments.

National implementation of climate change net-zero goals requires oversight and supervision. Markets and actors will not act left to their own devices, absent regulation and reporting. NCBs or councils are needed to press the rate of change and measure annual plans and performance against stated goals, to supervise and oversee application and compliance with stated goals. As governments delegate technocratic authority on carbon market oversight to organizations better suited to day-to-day supervision and oversight, they can turn to the equally important tasks around speeding technology innovation, diffusion, and disruption.

On speeding the rate of innovation and diffusion

We have very little time to reach our decarbonized goal in 2050. As I have stressed, governments, collectively and nationally, are essential actors. They must, via green industrial policies, support new technologies as they emerge, pull towards us, and steepen the technology diffusion S-curve of adoption. We can see this is working in the utility sector and renewables, which are now price competitive with fossil fuels. We can see this also in the development and adoption of EVs, which are set to accelerate literally and figuratively in the 2020s. We can see it in the rise of battery technology efficiency and falling pricing. In these sectors, we can see sustained government support, altered incentives, a steepening of the S-curve, continued innovation, and price cuts. These new engines of economic growth illuminate the paucity of denialists' arguments. Costs of new technologies are not static or too high. Rather, they have rapidly fallen. Innovation has not slowed but instead continues. Technological innovation is dynamic, disruptive, ongoing, and iterative. Governments have a key role to play in this shift to and adoption of green technologies.

There are still huge diffusion and application challenges ahead, from agriculture to construction, to industrial production, to airlines and shipping. In 2021, we have barely begun the needed industrial transition. Governments must speed the pace of change and innovation through altered regulation, incentives, support, penalties, and phase-outs. We know that markets alone and unsupervised cannot deliver on climate change net-zero goals. Governments must continue to set the guardrails and the glidepath, the mileposts and the measurements of our progress.

Within this regulated space, markets and firms will amplify the rate of the transition. In sector upon sector, leading firms are already doing so, grasping the challenge, recognizing the societal and business imperative of climate change mitigation and net-zero goals. First movers are already seeing the positive effects of their farsightedness in terms of growing market share, leaping equity values, and business prospects for the future. As many markets and increasing numbers of firms reach the climate change narrative tipping point, so too are communities and voters making the leap.

Our climate story is changing

In 2021, we are at a narrative tipping point in many regions and communities across the globe. This potentially crucial shift is being driven by demographics, as young, environmentally aware workers and investors take charge and drive market changes. In response, action on climate change is being demanded by asset managers, bankers, financiers, and CEOs of forward-focused firms, who hear investors' demands and see the contours of Green Globalization 2.0 taking shape and who want to lead, not lag. A shift in our narratives and conversations is also being demanded by activist groups, including the Fridays for Freedom movement led by Greta Thunberg, the

Extinction Rebellion nonviolent actions in cities across the world, and numerous others. Action on climate change is being increasingly demanded by citizen voters.

A poll taken in 2021 demonstrates the shift in our climate stories. The poll, the largest ever conducted on climate change, included 550,000 people across 50 countries, half of whom were between 14 and 18 years old. Across all countries, 64 percent of participants viewed climate change as an emergency requiring urgent responses from countries. In the UK and Italy, 81 percent polled held this view. In the US, 65 percent agreed climate change was an emergency, a notable and positive finding (BBC, 2021). Voters today see climate change happening around them and are alarmed. They want action.

Government leaders must stop being afraid of a backlash to radical climate change policy action and seize upon and support the narrative shifts that are already underway and heed voter demands. Leaders in Glasgow should simultaneously do more to foster fact-based storytelling on climate change and the options ahead.

Let's talk about it

The climate change stories we tell ourselves and that we use to understand its complexity vary. Our climate conversations must be ongoing, civic, face-to-face, and based upon a common understanding of the facts that underpin the need for action. In many locations, the facts of climate change are not in dispute. They are visible in the forest fires of Australia, Siberia, and California, and in the flooded landscape of Bangladesh. However, in some communities, the climate change crisis is still disputed.

Achieving breakthroughs in climate crisis economics, and in the implementation of policies designed to get us to net zero before we trigger climate tipping points, requires that we understand and agree the facts upon which policy action can then be constructed. Properly facilitated, such conversations can form the basis of a renewed, enlarged consensus on climate change and our responses. These conversations should be reasoned, depoliticized, and grounded in the scientific data, which are indisputable. We know that once communities agree on the facts, action and policy solutions can then be discussed and agreed and a consensus shift achieved. I have shown how this can play out, with evolution from denial to listening, to dialogue, to agreement, and to action. This needs leadership. It is not sexy. It is repetitive. It is necessary. Such conversations must take place at multiple levels across all communities.

Can such old-fashioned civic conversations match the power of internet memes, conspiracies, and 'alternative facts'? In an atomized, digitized world where many individuals operate illuminated only by the screens of their personal electronic rabbit holes of disinformation, calls for face-to-face conversations on the climate crisis seem quaint, perhaps anachronistic. Nonetheless, community conversations can help bridge the divide within communities and help re-establish a sense of commonality of facts and the need to act at the local and regional level.

It is in our communities where we directly affect climate change, through changes in planning, city design, renewables and electrification, transportation, housing, and construction, on our streets, in our parks, and through our personal practices. It is in our local communities that these climate change conversations can have maximal impact.

On devolving to deliver for the climate

I have stressed that national goals are essential, as are altered incentives, pricing signals, and regulation of markets and practices. This is not enough. Power and authority to act, together with the resources to do so, should be devolved to communities that must take ownership of the climate responses and transition glidepaths. If we are to achieve net zero and strengthen the local and lived stories needed to secure it, communities, regions, and towns must play their role, understand climate change, help fashion our responses, and own them. For when people own something and feel they have a stake in the outcome, they can bear greater burdens, do more to achieve the common goal, and change their personal conduct. Towns and regions are taking this forward, supporting conversations, shifting plans. Today, many localities (but not yet enough) are on net-zero journeys, redesigning and regreening neighbourhoods and making their communities more sustainable, liveable, and productive. As local communities change their environments for the better, the green industrial transformation will foster economic growth and progress.

On constructing Green Globalization 2.0

Green Globalization 2.0 can deliver broader-based growth, more skilled jobs, an expanding working population, higher wages, higher productivity, and an end to secular stagnation. Robert Gordon's (2016) end of innovation, and Lawrence Summers' (2014) secular stagnation can both be addressed and perhaps solved by the process of industrial transformation that is beginning to take place and that must accelerate. Constructing, rebuilding, and redesigning the 50 shades of green in our renewable electrified future will be localized, real, and industrial. Green Globalization 2.0 is an economy-wide, decades-long, deep, and broad real economy transformation. It will involve us all and affect us all. The breadth and depth of the transformation means it can power our economies and societies and help us reimagine and fashion them.

President Biden grasps this and wants to supercharge US plans to respond to the danger posed by climate change. He understands this will be the engine of growth going forward. He states:

> We know what to do, we just have got to do it … we are dealing with this existential threat. In dealing with it we can assure our future growth and prosperity … putting millions of American to work in good paying union jobs.
>
> *Biden, 2021*

The US president's activist climate change narrative is one of dynamic growth and opportunity, not of degrowth or exclusively of danger and threats. Biden's plans for America's greening will begin to align the world's largest economy to the task of achieving net zero. This governmental shift will support the transition and accelerate it via policy action, regulation, incentives, taxation, and market shifts. Green Globalization 2.0 in the US is a process, not an event. It will be resisted in some quarters – by the laggards and polluters, and by those who stand to lose out as we ensure a sustainable tomorrow. But the momentum is increasingly with those governments, businesses, activists, and voters seeking to address climate change and alter our stories. These leaders understand that issues of fairness and economic equity matter to the outcome.

On fairness and equity

As humans, we demand fairness. We require it in our childhoods and in our adult interactions. We reject those who break the rules repeatedly, and those who take advantage and refuse to cooperate. We also require fairness and equity in confronting and addressing climate change in the decades ahead. Governments need to recognize the fairness imperative and ensure a just transition within and across countries. This is not some socialist concept of forced equality. What is required is modestly better burden sharing of climate change costs and risks as well as sharing of the benefits of the transition. There can be no freeloaders.

A just transition internationally is also possible if wealthy governments raise their monetary support. I recommend a doubling of the annual support from advanced countries from US$100 billion per year to US$200 billion. Polls repeatedly show that advanced economy populations both overestimate the paltry amount of aid currently given but also think we should be more generous. This altruistic public stance provides an opening to act, to secure agreement from all states to net zero in 2021 and beyond, and to speed the transition among lower-income countries.

Sadly, this type of modest further investment and resource transfer is unlikely to happen. Advanced countries are loath to increase support, even as they spend US$14 trillion on fiscal measures to address the pandemic in just over 12 months. Nonetheless, leaders need to recognize that a refusal to support the transition of lower-income countries to net zero threatens to undermine common COP net-zero goals and the global commons. Sensible, inexpensive, multilateral, modest investments today will help ensure we secure the common goal and a sustainable future tomorrow. These investments can increase growth rates and benefit populations in both the emerging and advanced world.

Refusing to act collectively on climate change and in support of a just transition could undermine our net-zero goals, potentially leading to greater instability in fragile states and regions, and foster state collapse and social, political, and economic crises, and human tragedies. As Carney (2020) observes, 'you cannot self-isolate from climate change'. We will not be able to isolate ourselves from the ill effects of a failure to support a just transition today when climate crises and disasters strike tomorrow in Africa, Central America, or elsewhere, driving people across borders in search of survival.

Change is coming

This is a race of tipping points that we must win to ensure human and non-human survival. Our stories are changing, and our narratives are evolving rapidly. We must get to the tipping point on a global consensus on action before we hit climate tipping points of no return. Incrementalism must be rejected. We need new stories – new understandings – upon which we can construct a green, sustainable tomorrow. Many have already begun this reimagining. I believe we are very close to a narrative tipping point among governments, markets, and firms, and among voters, across regions, and in our communities and cities. Change is coming, and the faster the better. We need to all do our part to speed us towards this common planetary goal.

References

BBC. (2021) 'Climate change: Biggest global poll supports "global emergency"', 27 January. [Online]. Available at: www.bbc.com/news/science-environment-55802902 (accessed: 24 May 2021).

Biden, J. (2021) 'Remarks by President Biden before signing executive actions on tackling climate change, creating jobs, and restoring scientific integrity', White House, Washington, DC, 27 January [Online]. Available at: www.whitehouse.gov/briefing-room/speeches-remarks/2021/01/27/remarks-by-president-biden-before-signing-executive-actions-on-tackling-climate-change-creating-jobs-and-restoring-scientific-integrity (accessed: 19 February 2021).

Carney, M. (2020) BBC Reith Lectures. Lecture 4 [Online]. Available at: www.bbc.co.uk/programmes/articles/43GjCh72bxWVSqSB84ZDJw0/reith-lectures-2020-how-we-get-what-we-value (accessed: 14 January 2020).

Gordon. R. (2016) *The Rise and Fall of American Growth*. Princeton: Princeton University Press.

Kerry, J. (2021) 'Opening statement at Climate Adaptation Summit 2021', Rotterdam, the Netherlands, 25 January [Online]. Available at: www.state.gov/opening-statement-at-climate-adaptation-summit-2021 (accessed: 19 February 2021).

Mann, M. (2019) Interview, Yale Climate Connections. The Yale Center for Environmental Communication, Yale School of the Environment, Yale University, New Haven, Connecticut.

POSTSCRIPT

Change happens slowly and then fast

After the completion of this book in March of 2021, the rate of narrative and climate change policy shift accelerated and became more ambitious. The US administration called a virtual climate summit on April 23, at which President Biden committed the country to cutting GHG emissions by 50–52 percent by 2030 compared to 2005 levels; a dramatic and significant commitment, the implementation of which will affect the entire US economy. At that summit the US and China also stressed they would work together to achieve climate goals.

This ambitious American commitment, and others by allies, are altering and speeding policy decisions. There is a race to seize green opportunities out of a climate crisis. A competitive tension is seen between the US and China, which is spurring each side to reach higher and push faster towards common and national green economic goals. This is what we would hope to see.

For Americans, the challenge now centres on implementation and oversight, coupled to financial and regulatory policy shifts. Here Biden's ambition crashes against reluctance in Congress, and a Senate in which he has a majority of one (the vote of the vice-president). Can the president convert promise into actuality? It remains to be seen. Biden is shifting regulations fast, changing incentives, altering policy direction. But the effectiveness of this will be limited if the president fails to pass fiscal elements of his green industrial policy.

China too faces a challenge of huge proportions: to rapidly redirect, reengineer, and realign the second largest economy on the planet to a green and greener destination while continuing to spur broad-based growth and prosperity. China has further to go, on GHG reductions, than the US. But President Xi Jinping can direct the country's economy in ways impossible for an American president, and initial indications are that this wholesale reorientation has begun. In other words the policy shift is now getting underway.

DOI: 10.4324/9781003037088-11

What we see is a race between a green American modernization reimagination and reindustrialization and the transformation of state capitalism with Chinese characteristics into a green PRC.

This increased ambition among leading global competitors bodes well for COP26. The dynamic means laggard states are coming under much greater pressure to promise and deliver more (whether they are Russia, Brazil, Australia, or others).

These big public leaps in the US and China mean that COP26 can focus more on delivery, enforcement, pricing mechanisms, and monitoring mechanisms that can shift commitments into facts on the ground, across the globe.

This is good news. In the end climate change commitments only matter if they can be transformed into altered outcomes, GHG emission reductions, market sentiments, expectations, and a greening of business strategy and decisions.

The shift already visible by May 2021 gives some reason for optimism that the world may indeed be reaching a narrative and climate change policy tipping point – finally. Change, which so often happens too slowly, now appears to be happening fast. If this is the case, we can more likely avoid climate-driven tipping points of no return.

A net-zero transition is possible and achievable. We can do it. Let us get on with it.

INDEX